Geological Society Professional Handbooks

Stratigraphical Procedure

Geological Society Professional Handbooks
Society Book Editors

It is recommended that reference to this book should be made in following way:

Rawson, P.F., Allen, P.M., Brenchley, P.J., Cope, J.C.W., Gale, A.S., Evans, J.A., Gibbard, P.L., Gregory, F.J., Hailwood, E.A., Hesselbo, S.P., Knox, R.W.O'B., Marshall, J.E.A., Oates, M., Riley, N.J., Smith, A.G., Trewin, N. & Zalasiewicz, J.A. 2002. *Stratigraphical Procedure*. Geological Society, London, Professional Handbook.

GEOLOGICAL SOCIETY PROFESSIONAL HANDBOOK

Stratigraphical Procedure

BY

P. F. Rawson, P. M. Allen, P.J. Brenchley, J.C.W. Cope, A.S. Gale,
J. A. Evans, P.L. Gibbard, F. J. Gregory, E.A. Hailwood,
S.P. Hesselbo, R.W.O'B. Knox, J.E.A. Marshall, M. Oates, N.J. Riley,
A.G. Smith, N. Trewin & J.A. Zalasiewicz

2002
Published by
The Geological Society
London

THE GEOLOGICAL SOCIETY

The Geological Society of London (GSL) was founded in 1807. It is the oldest national geological society in the world and the largest in Europe. It was incorporated under Royal Charter in 1825 and is Registered Charity 210161.

The Society is the UK national learned and professional society for geology with a worldwide Fellowship (FGS) of 9000. The Society has the power to confer Chartered status on suitably qualified Fellows, and about 2000 of the Fellowship carry the title (CGeol). Chartered Geologists may also obtain the equivalent European title, European Geologist (EurGeol). One fifth of the Society's fellowship resides outside the UK. To find out more about the Society, log on to www.geolsoc.org.uk.

The Geological Society Publishing House (Bath, UK) produces the Society's international journals and books, and acts as European distributor for selected publications of the American Association of Petroleum Geologists (AAPG), the American Geological Institute (AGI), the Indonesian Petroleum Association (IPA), the Geological Society of America (GSA), the Society for Sedimentary Geology (SEPM) and the Geologists' Association (GA). Joint marketing agreements ensure that GSL Fellows may purchase these societies' publications at a discount. The Society's online bookshop (accessible from www.geolsoc.org.uk) offers secure book purchasing with your credit or debit card.

To find out about joining the Society and benefiting from substantial discounts on publications of GSL and other societies worldwide, consult www.geolsoc.org.uk, or contact the Fellowship Department at: The Geological Society, Burlington House, Piccadilly, London W1J 0BG: Tel. +44 (0)20 7434 9944; Fax +44 (0)20 7439 8975; E-mail: enquiries@geolsoc.org.uk.

For information about the Society's meetings, consult *Events* on www.geolsoc.org.uk. To find out more about the Society's Corporate Affiliates Scheme, write to enquiries@geolsoc.org.uk.

Published by The Geological Society from:
The Geological Society Publishing House
Unit 7, Brassmill Enterprise Centre
Brassmill Lane
Bath BA1 3JN, UK

(*Orders*: Tel. +44 (0)1225 445046
Fax +44 (0)1225 442836)
Online bookshop: http://bookshop.geolsoc.org.uk

British Library Cataloguing in Publication Data
A catalogue record for this book is available from the British Library.

ISBN 1-86239-094-0.

Typeset by Bath Typesetting, Bath, UK
Printed by The Alden Press, Oxford, UK.

Distributors
USA
AAPG Bookstore
PO Box 979
Tulsa
OK 74101-0979
USA
Orders: Tel. + 1 918 584-2555
Fax + 1 918 560-2652
E-mail bookstore@aapg.org

Australia
Australian Mineral Foundation Bookshop
63 Conyngham Street
Glenside
South Australia 5065
Australia
Orders: Tel. +61 88 379-0444
Fax +61 88 379-4634
E-mail bookshop@amf.com.au

India
Affiliated East-West Press PVT Ltd
G-1/16 Ansari Road, Daryaganj,
New Delhi 110 002
India
Orders: Tel. +91 11 327-9113
Fax +91 11 326-0538
E-mail affiliat@nda.vsnl.net.in

Japan
Kanda Book Trading Co.
Cityhouse Tama 204
Tsurumaki 1-3-10
Tama-shi
Tokyo 206-0034
Japan
Orders: Tel. +81 (0)423 57-7650
Fax +81 (0)423 57-7651

Contents

1. Introduction

Stratigraphy lies at the heart of the geological sciences, for it provides the fundamental framework on which depends our interpretation of all aspects of Earth's evolution. Stratigraphical procedure and practice date from the beginnings of our science, but have evolved to embrace an ever-increasing number of approaches. However, the essential purpose of stratigraphy as a whole remains simple. It is to create an ordered history of Earth, and of the physical, chemical and biological processes that take place on it and within it. For example, in the palaeogeographical reconstruction of a sea floor, it is important that the various components – submarine fans, shelf mud blankets, beach and tidal flat sands, etc. – are correctly ordered in time as well as in space. If they are not, then our interpretation becomes seriously flawed, as do any hypotheses regarding cause and effect in the geological past. Thus, it is of primary importance to understand the relative age of rocks, fossils and processes, that is to have reliable correlation.

The diverse areas of stratigraphy are essentially concerned with:

- Establishing the geometrical shape and spatial relationships of rock bodies, to define **lithostratigraphical** and **lithodemic units**
- Calibrating and correlating successions with one another and with the chronostratigraphical standard. For the Phanerozoic, the main tool has been **biostratigraphy** – the use of fossils in calibration and correlation. But many other methods are also utilized, as shown below
- Establishing a formal (relative) time framework – **chronostratigraphy**. This is ultimately defined by the placing of 'golden spikes' at carefully-chosen levels within rock successions
- Determining numerical ages for units, i.e. **geochronometry**
- Synthesizing the different approaches to produce the most refined correlation possible, i.e. **holostratigraphy**.

This Handbook, produced by members of the Stratigraphy Commission of the Geological Society, is designed to explain the different stratigraphical methods, show how they can be applied by the practising geologist and, where appropriate, indicate their current limitations. It is not a formalized stratigraphical code; the reader can turn to the *International Stratigraphic Guide* (Salvador 1994) or the *North American Stratigraphic Code* (North American Commission on Stratigraphic Nomenclature 1983) for that. But it does try to encourage a common approach to stratigraphical practice by offering clear guidelines on usage. It is aimed primarily at geologists working in the United Kingdom and therefore mainly utilizes UK examples. It also draws attention to the main UK stratigraphical databases.

Inevitably, this edition has drawn on previous ones. The Geological Society's then Stratigraphy Committee first produced a provisional stratigraphical code over 30 years ago (George *et al.* 1967), quickly followed by the more detailed *Recommendations on stratigraphical usage* (George *et al.* 1969). The latter was revised by Harland *et al.* (1972) to form *A concise guide to stratigraphical procedure*. Later versions (Holland *et al.* 1978;

Whittaker *et al.* 1991) dropped 'concise'! That reflected the development of new approaches and methods of correlation that have revitalized stratigraphy. Taken together, they lead to a holostratigraphical approach which is providing increasingly precise correlation for most Precambrian and Phanerozoic successions.

2. Establishing the rock succession

2.1. Lithostratigraphical units

Lithostratigraphical units are sedimentary or igneous units that conform to the Law of Superposition. **Lithostratigraphy** embraces the description, definition and naming of these units.

2.1.1. Description of lithostratigraphical units

When a new unit is defined, or an existing one formally revised, the following should be described:
- Lithological characteristics, including petrography, mineralogy, geochemistry, fossil content and lateral variation.
- Relationship with stratigraphically adjacent units, both vertical and lateral.
- Nature of boundaries: whether gradational or sharp, conformable or unconformable. In particular, describe how the base can be recognized.
- Thickness at the type section and its lateral variation.
- Exact location of the type section, using eight-figure grid references or, where appropriate, the Global Positioning System (GPS).

2.1.2. Definition of lithostratigraphical units

Units are defined within a hierarchical framework:

Supergroup
Group
Formation
Member
Bed

Whatever the scale, the defined unit should have, or originally have had, lateral continuity.

The basic unit is the **formation**, which is generally defined as the smallest mappable unit. However, 'mappability' is a loose criterion, for it depends on the scale of mapping. In UK practice, a formation should be mappable and readily represented on a 1 : 50 000 map scale: individual members may be mappable at this scale, but are not necessarily so. Formations generally vary in maximum thickness from a few metres to several hundred metres.

A formation has lithological characteristics that distinguish it from adjacent formations. However, it is not necessarily lithologically homogeneous – for example, the Cleveland Ironstone Formation in the Lower Jurassic of Yorkshire includes several coarsening-upward cycles stacked one upon another. Furthermore, there may be some quite distinctive sub-units that are best regarded as members. For example, from Dorset to Yorkshire the Kimmeridge Clay Formation (Jurassic) is usually argillaceous throughout, but in south Dorset it contains a local ironstone, the Abbotsbury Ironstone Member, and in north Lincolnshire a local sandstone, the Elsham Sandstone Member.

The boundaries of a formation may be sharp, interdigitating or gradational. In a gradational sequence they may have to be fixed arbitrarily, for example where increasing grain size reaches a particular grade, or at the base of a distinctive marker bed, or where there is a significant change in bed thickness. Interdigitation provides particular problems in nomenclature. A typical example is seen on the BGS Snowdon 1 : 50 000 geological sheet (No. 119) where the Ordovician Moelwyn Volcanic Formation comprises three members at the edge of its outcrop. The upper and lower are volcanic members, the middle comprises siltstones and sandstones which are indistinguishable from the main body of the host Nant Ffrancon Subgroup, and are not attributed to the Moelwyn Volcanic Formation on the map along strike where the volcanic members are missing. Effectively, this means that the formation cannot be shown on the map except where it is wholly volcanic. Elsewhere, its upper and lower members are enclosed by the Nant Ffrancon Subgroup, rather than within their host formation. This contravenes the rules in the *code*, but inevitably such pragmatic solutions are often necessary.

The Kimmeridge Clay Formation exemplifies the problem of how widely a name should be used when former lateral continuity is uncertain. The formation extends from eastern England into the southern North Sea and similar mudrocks are preserved as far north as the northern North Sea. Original lateral continuity is likely (Cope & Rawson, map J10 *in* Cope, Ingham & Rawson 1992), but while the name is used throughout the UK sector of the North Sea it is replaced in Norwegian waters by the Draupne and Mandal Formations (Vollset & Doré 1984).

A formation may stand alone or it may be linked with contiguous formations into a **group**. Definition of a group may be contentious. Some authors look for common lithological characteristics, e.g. the Lias Group is dominantly argillaceous but contains limestone-, sandstone- and ironstone-dominated formations. Others focus on the genesis of the group and thus place less emphasis on lithological characteristics and more emphasis on changes in gross depositional and tectonic character. In the latter case, the bases of successive groups are commonly defined by basin-wide unconformities and their correlative conformities (e.g. Woodcock 1990) (see also *sequence stratigraphy*). A **supergroup** embraces contiguous groups and may also contain formations that have not been assigned to a particular group.

A **member** is a subdivision of a formation, but formations are not necessarily divided either wholly or partially into members. Members commonly occur in the marginal areas of a formation, representing, for example, a marginal sandbody prograding into a predominantly argillaceous basin.

The smallest formal unit is a **bed**. In lithologically monotonous sequences like the Chalk Group, formal bed names for distinctive fossil, marl and flint bands may have considerable reference value (e.g. Wood & Smith 1978; Fig. 2.1). In more variable sequences it is rarely worth giving a formal name to every individual bed, but there may be distinctive beds which can be traced over long distances that merit naming. If the bed

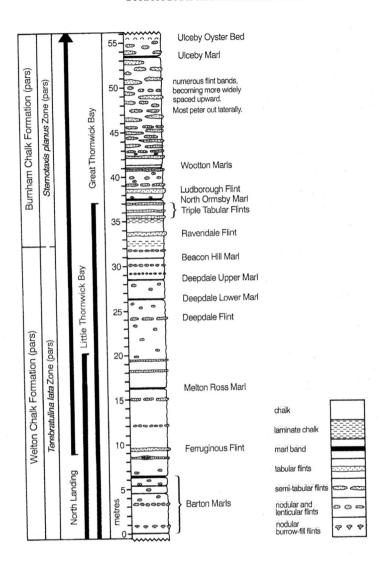

Fig. 2.1. Lithic log of part of the Chalk Group at Thornwick Bay and North Landing, Flamborough, North Yorkshire, showing named flint and marl bands (Reproduced from Rawson & Wright 2000, fig. 42, with permission from The Geologists' Association).

is named after a fossil, the fossil name is not italicized (e.g. the Middle Jurassic Boueti Bed, named after the brachiopod *Goniorhynchia boueti*). If the fossil name subsequently changes, the name of the bed does not.

The term **subgroup**, though not in the formal hierarchy, has to be used on occasion. In the Ordovician succession of Snowdonia mentioned above, when mapping was completed and the nomenclature reviewed (Rushton & Howells 1998), the Nant Ffrancon Group was reclassified as a subgroup within the Ogwen Group, rather than the latter being regraded as a supergroup. And a recent meeting of Chalk specialists, organised by the Stratigraphy Commission, agreed that the Chalk Group should be divided into the Grey Chalk and White Chalk subgroups (Rawson *et al.* 2001; Fig. 2.2).

We recommend that the term **division** be retained only for informal use as a term without hierarchical connotations.

STAGE	SUB-GROUP	FORMATION	MEMBER
CAMPANIAN	White Chalk	Rowe Chalk	
		Flamborough Chalk	
SANTONIAN			
CONIACIAN		Burnham Chalk	
TURONIAN		Welton Chalk	
CENOMANIAN	Grey Chalk	Ferriby Chalk	Plenus Marls

Fig. 2.2. General subdivision of the Chalk Group of the Northern Province.

2.1.3. Naming of new lithostratigraphical units

Normally a new unit is named after a distinctive geographical feature/area, preferably at the type section. The same geographical name should not be used for more than one unit.

It is recommended that, where possible, a lithological descriptor is added, though this is not mandatory and may be inappropriate where more than one lithology occurs. The name also indicates the hierarchical level of the unit. All the initial letters of such a formally defined name are capitalized (e.g. Ravenscar Group, Welton Chalk Formation). Additional terms such as lower, middle and upper should be used only informally and thus without initial capitals.

2.1.4. Stability and usage of existing lithostratigraphical names

Long-established stratigraphical names may not fit readily with the principles noted above. Major revision of the stratigraphy of a previously described area may necessitate the introduction of new names. But we recommend that for the sake of continuity existing names be retained wherever possible, albeit modified to fit into the modern hierarchical framework. In practice, this process is already well established; for example, the Wenlock Limestone is now the Much Wenlock Limestone Formation, the Lower Greensand is now the Lower Greensand Group and the Lias is now the Lias Group. Where irregularly-formed names, such as the Leziate Beds (part of the Sandringham Sands Formation) are assigned a position in the modern hierarchy then 'Beds' should be dropped, i.e. one would refer to the Leziate Member not the Leziate Beds Member. Other irregularly formed names can be conserved as long as confusion does not arise (e.g. MacGregor Marine Band).

Any formal lithostratigraphical name should normally be written in full. But where the name has three components (geographical, lithological and hierarchical), the lithological epithet can be dropped in abbreviation, following an initial citation of the full name. Where new work demonstrates that the hierarchical part of the name is inappropriate then its rank can be changed; for example the Brent sands of the northern North Sea were originally described as the Brent Formation but they have been reclassified subsequently as the Brent Group - which is now divided into five formations.

The nomenclature used for Carboniferous cyclothemic successions on British Geological Survey (BGS) maps follows conventions that differ significantly from formal lithostratigraphical practice. While the term 'group' has been used locally, for example for the Upper Border Group of Viséan age in the Northumberland Basin, it may equate either to a group or a formation in modern terms, and it was never further subdivided into contiguous units. The convention followed was to name significant limestones or shell beds in parts of the succession where they were present and sandstones where they were not. Coal seams and marine bands were named throughout. The argillaceous rocks between them were not attributed to any named unit. The succession as a whole was divided by colour on the maps into chronostratigraphical stages, each defined by a basal marine band or the nearest mappable sandstone or limestone to it. An attempt was made in the 1970s to divide some parts of the Upper Coal Measures of West Yorkshire into four, contiguous, named divisions. These are now regarded as informal units of member status and the use of the term division is discontinued.

2.2. Lithodemic units

Bodies of rock which do not conform to the Law of Superposition are described as lithodemic. They are generally composed of intrusive, highly deformed or metamorphic rocks, determined and delimited on the basis of rock characteristics. Their boundaries may be sedimentary, intrusive, extrusive, tectonic or metamorphic. A formal classification of lithodemic units was presented in the 1983 *North American Stratigraphic Code*, comprising a **lithodeme**, which is comparable to a formation, a **suite**, which is roughly equivalent to a group, and a **supersuite**, comparable with a supergroup.

The 1994 *International Stratigraphic Guide* does not recognize lithodemic units, but regards intrusive igneous bodies and non-layered metamorphic rocks of undetermined origin as special cases within lithostratigraphy. That guide advises against using the term suite. The term **complex** (see below) is used loosely. This approach provides little help to geologists mapping in complicated basement and plutonic terrains. It is recommended here that the procedures advocated in the *North American Stratigraphic Code* are followed, with the modifications noted below.

2.2.1. Definition of lithodemic units

This guide recognizes the following, two-fold hierarchy:

Supersuite	*Supercomplex*
Suite	*Complex*
Lithodeme	

The **lithodeme** is the fundamental unit. No lower rank units have been defined. It should possess distinctive lithological features and be internally consistent by comprising either a single rock type or a mixture of two or more types that serve to distinguish the unit from those bounding it. Normally a lithodemic unit should be given a geographical name combined with either a rock name or some term descriptive of its form. Examples are Shap Granite, Parwyd Gneisses and Arisdale Inlier.

A **suite** consists of two or more lithodemes of the same genetic class (igneous, metamorphic or sedimentary): it may refer to, say, a metamorphic suite or a plutonic suite. The term has been applied infrequently in its modern sense in Britain, though it is well established in the literature applied to plutonic rocks. Whitten (1991) gives a full account of the range of uses of the term and concludes that a set of plutons can be divided into different suites according to which petrographical or chemical characteristics are used to define the suite. It would seem, from this, to be undesirable to attempt any more precise a definition of suite than that provided by the *North American Stratigraphic Code*.

Problems of nomenclature may occur in zoned plutons, such as the Loch Doon Pluton in Scotland, which ranges compositionally from diorite to granite. If there is no need to recognize each component as a lithodeme, the pluton as a whole can be regarded as a lithodeme and its components recognized as lower rank, informal units, if necessary. However, an alternative approach, recommended here, would be to use the informal term pluton intermediate between lithodeme and suite. Thus the Loch Doon Pluton and various so-called complexes, such as the Sarn Complex on the Lleyn Peninsula and the Ratagain Plutonic Complex, would all be designated as plutons. This would leave the word suite to be used for a set of plutons as, for example, the Argyll Suite in the NW Highlands.

In the *North American Stratigraphic Code*, the use of the term **complex** is confined to an assemblage or mixture of rocks of two or more genetic classes. The rank of the term was not defined, but was said to be comparable to a suite or supersuite. In practice this can lead to problems, such as nested complexes; for example the Laxfordian Complex within the Lewisian Complex of NW Scotland. Where recent, practical attempts to apply lithodemic nomenclature have been made in North America the tendency has been to confine the term complex to a unit that has the same rank as a suite and consists of lithodemes of more than one genetic class. For higher rank units, the term **supersuite** is available. A supersuite has been defined as consisting of two or more suites or complexes, having a degree of natural relationship with each other, but not necessarily of the same genetic class. By implication a supersuite could consist entirely of complexes. However, there appears to be a resistance in the UK to the latter usage. As a pragmatic solution we recommend the introduction of the term **supercomplex** for such as the Lewisian, thus limiting supersuite to a grouping of two or more suites.

2.2.2. Special cases

Two special cases have been defined. Where rocks have been brought together in complicated relationships by tectonic processes the unit may be referred to as a **structural complex** whether or not the components are of different genetic classes. The utility of this term is doubtful except where it is used specifically for a collection of different elements brought together by structural processes; for example, tectonic mélanges where different rock types have been tectonically mixed together. When the different elements have merely been dislocated by faulting, such as the Highland Border Complex, it is a complex *sensu stricto*, not a structural complex.

The second special case is a **volcanic complex** which is characterized by a diverse assemblage of extrusive rocks, related intrusions and their weathering and erosional products.

Because a mixture of genetic classes is required in a complex there may be components which can be divided lithostratigraphically as well as lithodemically. A typical volcanic complex, for example, may consist of rocks of all three classes. Extrusive volcanic and intercalated sedimentary rocks may be named following lithostratigraphical procedures, while the associated dykes, sills and subvolcanic plutons, vent intrusives, metamorphic and metasomatic rocks are lithodemic. An example is the Rum Central Complex consisting of intrusive and extrusive igneous rocks, pyroclastic deposits and enclosures of Lewisian metamorphic rocks.

2.2.3. Problems of usage

There are many more examples of inappropriate use of the term complex in British geology than correct usage, even according to the *North American Stratigraphic Code* definition. Confusion is likely to remain as long as the term is not ranked. Thus it is recommended that **the term 'complex' is used only for units equivalent in rank to suite.** Some problem cases reflect historical precedent. The Rum Central Complex, for example, consists of igneous rocks and enclosed bodies of Torridon Group sedimentary rocks. However, among its named components is the Rum Layered Suite, which comprises three lithodemes, the Central Intrusion, Western Layered Intrusion and Eastern Layered Intrusion. Logically, it should be named the Rum Central Super-complex, but the size of this assemblage of rocks is considerably smaller than the average supergroup, with which a supercomplex is comparable, and may reflect on the inadequacy of a three-rank lithodemic classification scheme. In general, it is better to confine the terms supersuite and supercomplex to units equivalent in scale to supergroup and use informal terms at intermediate levels where detailed work justifies it.

A large number of informal terms have been used to describe lithodemic units. Those, such as series, which have other defined and formal uses, should not be used. Others, such as zone, are acceptable at the rank of lithodeme, with appropriate descriptors. In general, informal terms which describe rock units that could otherwise be described as suites, complexes, supersuites and supercomplexes should be avoided.

2.3. Allostratigraphical units

An allostratigraphical unit is a body of sedimentary rock that is defined and identified only by its bounding discontinuities. Although the formal units **allomember, alloformation** and **allogroup** were explained in the *North American Stratigraphic Code*, in practice, allostratigraphy has not been welcomed in the UK research community, though it has been used in the USA (e.g. Morrison 1985). Allostratigraphic units may be defined to distinguish between superposed deposits (a succession of lithologically similar alluvial and lacustrine deposits in a rift valley, separated by soil horizons), contiguous discontinuity-bounded deposits (onlapping lobes of soliflucted material from different glaciations), and geographically separated bodies (river terrace deposits), all of similar lithology; or to distinguish as single units discontinuity-bounded bodies characterized by lithological heterogeneity. The principles of naming these units are the same as for lithostratigraphical units. The bounding discontinuities are described in the Code as unconformities, disconformities or the present-day land surface. Fault boundaries are excluded.

While sequence stratigraphy (see below) was also developed to classify unconformity-bounded units, allostratigraphy covers the special case of rock bodies of similar lithology separated by discontinuities. The *International Stratigraphic Code*, however, made no such distinction and recommended a hierarchy of formal terms based on the fundamental unit **synthem** for all unconformity-bounded units – i.e. it did not use alloformation, etc.

The most easily recognized allostratigraphical units are river terrace deposits. They rest unconformably on bedrock; their upper surfaces are the current land surface, and the deposits of the different terraces may be lithologically indistinguishable from each other. That they are temporally separate can normally be established by their spatial relationships and altitude.

On BGS maps terrace deposits are classified in terms of their form and origin. Individual terrace surfaces are identified by numbers, the first terrace being the lowest and youngest. In well studied areas, numbering may be replaced by named deposits and surfaces, following international practice. There is a full succession of named terrace surfaces and underlying deposits along the River Thames, for example. These names are not always the same; thus, the Harefield Terrace is developed on the Gerrards' Cross Gravel deposit. There is disagreement among various authors about the rank of the Thames terrace deposits, but all have been defined as lithostratigraphical units; allostratigraphical classification and nomenclature have been ignored (see Gibbard 1985).

Alluvial and lacustrine deposits in a rift environment, in which stacked sediment units of similar lithology are separated by disconformities, are allostratigraphical units. Note that these are the same as **depositional sequences** in sequence stratigraphy, or **subsynthems** in the classification of unconformity-bounded units given in the *International Stratigraphic Code*.

It is a requirement in lithostratigraphy adopted in both the North American and International stratigraphical codes for a succession of sedimentary rocks of uniform composition to constitute a single lithostratigraphical unit, even if there is an unconformity in the middle of it. An example is the Permian Exeter Group of red breccias and sandstones in Devon, in which unconformities separating the Cadbury Breccia, Thorverton Sandstone and Crediton Breccia cover a time interval of over 30 myr (Edwards & Scrivener 1999). That approach is questioned here. Unconformities and disconformities are legitimately used to define bounding surfaces of lithostratigraphical units, and even within the red breccias of the Exeter Group there are differences in clast content that can be used to distinguish the upper from lower breccias. Thus there may only be a need for an allostratigraphical scheme when it is impossible to distinguish the rocks above and below a discontinuity by any method. However, Quaternary researchers have no difficulty ascribing a unique identity to compositionally similar terrace gravel deposits using altitude and consider that this is sufficient justification to give individual names (or numbers) to deposits of the same composition. Disconformity-bounded units, separated by palaeosols, in the Tertiary Antrim Basalts and Devonian red-bed successions would also be regarded, strictly, as allostratigraphical units, but have been classified successfully as lithostratigraphical ones.

There appears to be a consensus that if superposed or contiguous deposits of similar composition can be distinguished from each other by any method, it is justifiable to classify them as separately named lithostratigraphical units, a practice which is endorsed in this guide.

2.4. Morphostratigraphical units

The unique nature of geological sequences in the Quaternary has meant that in many situations geomorphological approaches have been included in the subdivision of sediment and erosional sequences. The identification and mapping of land surfaces, either developed upon sediment bodies or in some cases on bedrock, has been used as a means of interpreting relative chronologies in many regions. Typical examples include glacial moraines and associated landforms, dunes, fossil shorelines and river terraces (but see also *Allostratigraphical units*).

In each of these cases the so-called morphostratigraphical unit is used to denote a body of sediment that is identified primarily from the surface form it displays (Frye & Willman 1962). Central to the recognition of such units is that they include both landform and lithology in their definition (Bowen 1978). Clearly these units are not directly comparable to standard lithostratigraphical units, where vertical and lateral changes, as well as relationships to other units, can generally be observed unambiguously. Morphostratigraphical units should, therefore, only be given informal status (Richmond 1959). However, in some Quaternary sequences, particularly in regions of recent glacial recession, moraine ridges mapped over considerable distances are often afforded virtually formal status, e.g. the Salpausselkä Moraines in southern Finland. Similarly, shorelines, either raised or submerged, have been used in a comparable sense in some regions.

Nevertheless, the apparently simple external morphology of some landforms, such as river terrace surfaces (see also *Allostratigraphical units*), dunes or ice-marginal formations, commonly masks internal complexities of sediment sequences preserved beneath or within them. For this reason, whilst morphostratigraphy might prove to be very useful in some regions, it should never be regarded as a substitute for, or a short-hand way of referring to, other more precise types of stratigraphical unit, such as outlined under *lithostratigraphical units*.

2.5. Tectonostratigraphical units

Tectonostratigraphy refers to the temporal and spatial relationships of rock units brought into juxtaposition by tectonic activity. There are no formal tectonostratigraphical terms.

In recent years the term **terrane**, spelled in the North American fashion, has become widely accepted to describe a segment of the crust bounded by major, usually transcurrent, faults. Where the area bounded by the faults has markedly different stratigraphy and tectonic history from neighbouring terranes and entirely unknown relationships to them it is called a **suspect terrane**. If it can be proved that the terrane has been displaced a large distance and is not indigenous to the cratonic margin to which it is now attached it is called an **exotic terrane**. The rock units within a terrane can usually be classified and named following lithostratigraphical and/or lithodemic procedures.

For smaller units, commonly within a terrane, there is some overlap with the lithodemic classification. For example, a complex comprising fault-dislocated rocks of different genetic classes, situated along a major lineament adjacent to a terrane boundary, is also a tectonostratigraphical unit. Another term, **division**, has been used to describe components of the Moine rocks of NW Scotland, in part bounded by major thrusts. Each division was divided into formations. These divisions have since been reclassified as groups within a Moine Supergroup. They contain an internally consistent stratigraphy appropriate for lithostratigraphical classification, despite tectonic disruption at the boundaries of the divisions. It is recommended that 'division' is not used, even informally, in this sense.

There is no single term available to describe the smallest mappable tectonostratigraphical unit, but the term **tract** has been used for the numbered divisions of the Gala Group turbidite succession in the Southern Uplands thrust belt (Stone 1995). Each tract is bounded by named faults. Lithologically, there is little to distinguish one from another and, in the absence of the Ordovician to Silurian Moffat Shale Group, which is finely divisible by graptolites, it is generally unusual to find evidence to allow formal identification or correlation of lithostratigraphical units. But locally, parts of some numbered tracts have been characterized as formations. In the overlying Hawick Group in the Southern Uplands, for instance, the fault-bounded tracts are sufficiently different from each other to have been classified and named lithostratigraphically even though it may not be possible to recognize the tops and bases of each unit.

3. Calibrating and correlating successions

Once individual successions have been defined, there are several ways in which they can be calibrated individually and then correlated with one another and with the chronostratigraphical scale.

3.1. Biostratigraphy

Biostratigraphy is the use of fossils in stratigraphy. It relies on the study of *in situ* fossil distributions to allow recognition of stratigraphically restricted and geographically widespread taxa or populations, which enables subdivision and correlation of lithostratigraphical successions. Such taxa may be selected as index fossils and used as the basis of biostratigraphical correlation – one of the stratigrapher's most powerful tools for correlating Phanerozoic sequences. The basic unit of biostratigraphy is the **biozone**, which is formally described in terms of its fossil indices and content. Biozones are then ordered in stratigraphical position ultimately to allow correlation of lithostratigraphical units. Biozones can be of any thickness or duration. They can be local to world-wide in scale.

Most Phanerozoic successions with fossil remains, from the lower Cambrian to the Recent, have been subdivided biostratigraphically. While some of the lower Cambrian non-trilobite zones may be several millions of years in duration (Landing *et al.* 1998), resolution is usually much finer, with a regional resolution of as little as 600 000 years per biozone (e.g. Jurassic ammonites: Cope *et al.* 1980). Biozones may be further divided. **Sub-biozones** are defined by taxa that may be of only local importance within the more regionally extensive hierarchy of biozones. Erection of one or more sub-biozones within a biozone does not mean that the whole of the biozone has to be thus subdivided. **Biohorizons** represent single, laterally widespread palaeontological events, which are not vertically extensive and often provide palaeontological evidence of maximum flooding surfaces or condensed units as applied in sequence stratigraphy (Armentrout & Clement 1991). An example of the use of biohorizons is illustrated by the ammonite-rich faunas developed within the Aalenian and Bajocian (Jurassic) of southern England (Callomon & Chandler 1990). Such local biostratigraphical resolution may identify units with a duration of less than 100 000 years (e.g. 43 000 for the Callovian; Cope 1993).

Historically, biozones were usually defined in areas of supposed 'continuous' sedimentary deposition under marine influences (see Newell 1967; Ager 1993, for a discussion of the importance of stratigraphical gaps in marine sequences). However, biozones are increasingly being established for lacustrine palaeoenvironments (e.g. using diatoms, spores and pollen) and terrestrial palaeoenvironments (e.g. using mammals; Shuey *et al.* 1978).

3.1.1. Naming biozones

Biozones should normally be named after their most characteristic fossil, using standard Linnaean binomial notation (generic form capitalized, specific name lower case, both names italicized, with 'biozone' capitalized, e.g. *Hyperlioceras discites* Biozone, or shortened using the specific name *discites* Biozone). If a binomial name is used for the biozone, it should be based on a validly published taxon and follow the relevant code of nomenclature (e.g. Zoological Code, Ride *et al.* 1985; Botanical Code, Greuter *et al.* 1988). If the name of the taxon then changes, so should the zonal name. Some authors also recommend that the type of biozone should be included in the name (e.g. *Didymograptus artus* Taxon-range Biozone), but this can be rather unwieldy. In practice it may be more useful to indicate the type of taxon used, e.g. the *Simbirskites speetonensis* Ammonite Biozone.

It must be noted that there are biozonation schemes in common use which do not follow this procedure but instead use alphanumeric notation. These include the P and N Tertiary planktonic foraminiferid schemes (Banner & Blow 1965; Berggren & Van Couvering 1974) and many commercially based biozonations, published or otherwise (e.g. MJ1-MJ25 Jurassic scheme of Partington *et al.* 1993a,b). Such schemes are too well entrenched to be replaced.

A reference section or sections should be presented for newly established zonal schemes, with fossil range charts and ideally, descriptive taxonomy or illustrative plates of biozonal taxa and possibly their associated assemblages.

3.1.2. Index fossils

Suitable fossil indices should be geographically widespread, common, stratigraphically restricted and morphologically distinct enough to enable unambiguous recognition. If possible, the taxa selected should be planktonic or nektonic (e.g. graptolites, ammonoids, planktonic foraminifera) to lessen the effects of facies-controlled distribution which limits all living organisms, but which can especially influence the distribution and dispersal of benthic taxa (e.g. most trilobites, brachiopods, benthic foraminifera). Most long-established bizonations have relied on macrofossils, especially trilobites, graptolites and ammonoids, but oil company drilling activities have led to the increasing use of microfossil groups. Principal microfossil groups studied extensively over the last 30 years include palynomorphs (acritarchs, spores, pollen, dinoflagellates), foraminifera (planktonic and benthic), nannoplankton, radiolaria, marine diatoms, ostracods and conodonts.

These microfossil groups possess many of the attributes outlined for biozonal indices, but with two important additions that relate directly to their small size (generally less than 1mm). Firstly, they are commonly preserved in prodigious numbers (e.g. *Globigerina* oozes; Radiolarites) and secondly, drilling usually destroys macrofossils but microfossils are normally returned to surface in small rock fragments (ditch cuttings). Conversely, some microfossil taxa are longer ranging than macrofossils so that zonations based on them may be less precise than those based on macrofossils. However, some micropalaeontological, and more typically palynological, schemes for offshore exploration areas (e.g. UK North Sea Jurassic) are comparable in precision with well

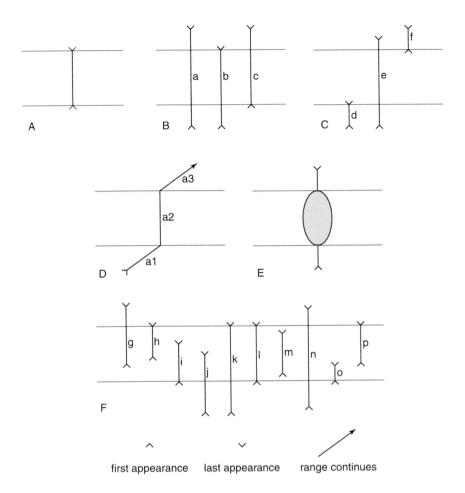

Fig. 3.1. Types of biozone. (A) Total Range (Local Range) Biozone; (B) Concurrent Range Biozone defined by overlap of taxa a, b, c; (C) Partial Range Biozone; (D) Consecutive Range Biozone (species a1, a2, and a3 are phylogenetically linked); (E) Acme Biozone; (F) Assemblage Biozone. (After Whittaker *et al.* 1996).

established macrofossil biozonation schemes. A good example of this is the correlation of onshore ammonite biozones with microfossil zonations for the UK Central North Sea Jurassic (Partington *et. al.* 1993a, 1993b). Copestake (1992) discusses the merits of various microfossil groups for biostratigraphy.

3.1.3. Types of biozone

There are several categories of biozone (Fig. 3.1), and the biozonation of a single succession can comprise a mixture of types.

Range Biozones reflect the actual range of taxa. A biozonal index does not have to occur throughout the entire vertical extent of its distribution. Hence there are several types of range biozone:

> *Total Range Biozone*: boundaries are defined on the combined first (evolution) and last (extinction) occurrences and represent the total stratigraphical and geographical range of the index taxon.
> *Local Range Biozone*: in practice, it is difficult to define total ranges and many so-called total range bizones are in fact local range biozones.
> *Concurrent Range Biozone*: defined on overlapping limits of stratigraphical ranges of two taxa (both tops and bases).
> *Partial Range Biozone*: established within the stratigraphical range of a taxon, with the biozone being actually limited by the appearance and/or extinction of other taxa (= Overlap Biozone of Hedberg 1976).
> *Consecutive Range Biozone* (= *Lineage Zone* of Salvador 1994): based on the range of a taxon within an evolving phylogenetic lineage.

Acme and **Assemblage biozones** are defined on population characteristics:

> *Acme Biozone*: the boundaries are defined by the maximum extent (abundance/ superabundance) of a particular taxon. The subjective nature (i.e. the differing assessments of relative abundance) of acme biozone definition often limits the recognition and regional extent of such units. Additionally, acmes are often related to microhabitat/localized nutrient influxes. Therefore, acme biozones tend not to be applicable over large distances.
> *Assemblage Biozone*: this is also subjective in definition and is characterized by a distinct assemblage or association of three or more species, which can be distinguished by its character from adjacent assemblages. This type of biozone is particularly influenced by facies/environment of deposition and can be difficult to define (Holland & Bassett 1988).

3.1.4. Boundaries of biozones

Biostratigraphy is a very practical application of the fossil record, so it is hardly surprising that there are two different approaches to the definition of biozonal boundaries.

In 'classic' biostratigraphy (Fig. 3.2a), where the biostratigraphical schemes are usually based on macrofossils collected at outcrop, biozones are normally defined by their base, marked by the first appearance (evolutionary or migratory) of a particular

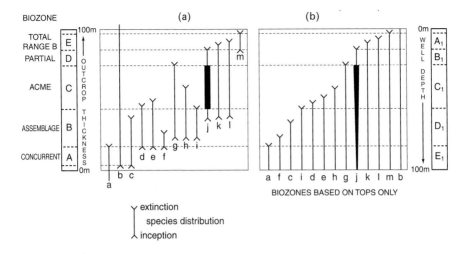

Fig. 3.2. Biozones from outcrop and borehole. Hypothetical comparison of Biozone construction from outcrop and commercial drilling methods based on the distribution of similar taxa. (a) Outcrop: Taxa distribution arranged according to evolutionary inception. (b) Well site: Taxa occurrence arranged according to final occurrence. Ranges extended due to downhole caving. From Gregory (1995).

fossil (which may not necessarily be the zonal index). The top is then defined by the base of the next overlying unit. In any given section the first appearance may change if subsequent discoveries extend known fossil ranges, so that one cannot apply a 'Golden Spike' approach to biozonal boundaries (e.g. Salvador 1994; or see Ager 1993 for a partisan approach).

Ranges of individual taxa are often indicated by reference to **first (evolutionary) occurrences (FO)** and **last (evolutionary) occurrences (LO)**. However, in the commercial world, drilling practices usually do not permit the confident selection of first occurrences (evolutionary inceptions), as downhole contamination of younger sediments (cavings) obscures these events (Fig. 3.2b). Therefore, the principal methodology, without recourse to expensive core or sidewall core retrieval, is to delimit biozones based on first downhole occurrences (usually extinction events) rather than evolutionary inceptions. Many acronyms have been suggested for these events, but it is recommended that the terms **First Downhole Occurrence (FDO)** and **Last Downhole Occurrence (LDO)** alone be used. The accurate resolution that such biozones provide can be used to biosteer horizontal holes through biozones during drilling (Shipp 1999).

Regional correlation of a biostratigraphical event may only approximate to a time-line due to diachronous taxa distribution. This is particularly evident with respect to the distribution and dispersal of benthic populations, which are often facies dependent. It can be supplemented by quantitative techniques, e.g. graphic correlation.

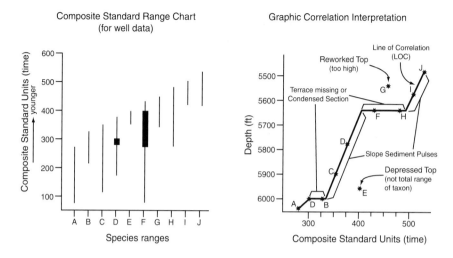

Fig. 3.3. Graphic correlation. Adapted from Neal (1996) to show construction of graphic correlation 2-axis graphs.

3.2. Graphic correlation

Graphic correlation is now becoming an important tool for examining the large datasets generated from offshore commercial drilling activities (e.g. Neal 1996). However, the potential of this technique has yet to be fully realised. As devised by Shaw (1964), graphic correlation used biostratigraphical data (measured position/depth) and simple two-axis graphs to optimize correlation, but it can be applied to all measured position/depth related information, for example, palaeomagnetic, lithological (e.g. correlatable events such as ash bands) and sequence stratigraphical boundaries. The method was used originally to compile taxa ranges from outcrop, and hence to determine relative sediment accumulation rates and the onset of sedimentation.

To use the technique, the most complete section is selected as a standard. Other, less complete sections are then included graphically by iteration to build up a composite section against which new sections can be plotted. A line of correlation (LOC) is then constructed by the stratigrapher, who judges the best fit of points (usually species ranges) when any new section is compared against the composite section (Fig. 3.3). This procedure can be used for large computerized datasets. However, in all of these interpretations it must be stressed that the palaeontologists' experience is paramount in the placement of the LOC.

When the sections are tied to chronostratigraphical frameworks, LOC patterns can be interpreted to indicate both true time relationships of events such as condensed/missing sections, which appear on the graph as 'terraces', and the depositional history, indicated by variations in the slope of the line (Fig. 3.3). Additionally, once total composited ranges have been constructed, the LOC provides a test for the total ranges of component taxa and can indicate the presence of reworked taxa as well as depressed ranges (i.e. not representative of the true total taxon range) (Fig. 3.3). Mann & Lane (1995) and Cooper & Lindholm (1990) are a good starting points for the methodology and use of graphic correlation.

3.3. Geophysical/petrophysical log stratigraphy

Boreholes permit investigation of the strata drilled using various geophysical methods, which allow the distinction and correlation of stratigraphical units and events. Geophysical logs record continuous digital wellbore petrophysical data, either whilst drilling (measurement while drilling – MWD, or logging while drilling – LWD) or after the borehole has been completed (wireline and coiled tubing logging). The digital data are generated by a suite of tools with an output as a series of depth-matched digital values that can also be displayed graphically, normally as linear traces. These data are integrated and used to create an interpreted lithological log, or occasionally an 'electrical facies' log, in which the petrophysical properties are the main discriminators. Interpreted lithology is usually corroborated at the wellsite or in subsequent analysis, from the fragments of rock returned to the surface as ditch cuttings in the drilling mud, from cores or from spot core samples (sidewall cores).

The commoner logging sensors used in modern drilling, often run as a combined wireline toolstring, are:

Resistivity (electrical resistivity). Measures the conductivity of the formation, from which water saturation, cementation, etc can be estimated accurately. Useful in identifying the presence of hydrocarbons.

Gamma ray log (GR). Measures natural gamma radiation, and is useful for discriminating between, for example, argillaceous facies and less radioactive arenaceous rock types, carbonates or evaporites.

Sonic log (δt). Measures the speed with which rock transmits acoustic energy (sound). Particularly useful in estimating porosity and identifying formations with characteristic signatures such as halite, which might not easily be preserved at surface.

Spontaneous potential log (SP). Measures the potential difference between the formation and the surface. Data from this tool is dependant on a difference in salinity between drilling mud and formation water. It is seldom used now as a primary lithology indicator, but can be useful in estimating formation water salinity.

Neutron log. Bombards the formation with neutrons and measures the gamma radiation returned from collision with hydrogen atoms (e.g. in formation water, oil). Used mainly in porosity determination, and consequently a good lithology tool in variable clastic sequences, particularly in conjunction with the density log.

Density log. Measures formation density, from which porosity can be calculated, and also discriminates rocks with characteristic uniform density such as evaporites and coal.

Dipmeter. Integrates four to eight radially spaced resistivity measurements to recognize surfaces crossing the wellbore (typically bedding) and to orientate their dip and strike. Unconformities can be recognized easily, but stratigraphically the tool is often used for the very fine resolution of its raw resistivity data, which can be used in particularly detailed correlation.

Image logs. Based on closely spaced, fine resolution resistivity or sonic log data, these devices provide a graphical image of the wellbore, from which bedding, induration, dip angle and direction, faults, and even sedimentary structure and trace fossils may be recognized.

Rider (1998) and Whittaker *et al.* (1985) detail the tools and interpretation of logs.

Continuous data recording of geophysical logs can have an advantage over lithological samples alone returned to surface, as the logs can add precision to depths, thickness, structural disposition and sedimentary succession. Other physical properties such as porosity and hydrocarbon saturation can also be determined in relation to the rocks drilled, although the vertical resolution of each logging tool has its limitations, and thin-bedded sequences will present 'smeared' results that tend to average the window being measured by the tool at any time during the logging run.

Once the lithological and petrophysical data have been interpreted, it is possible to define lithostratigraphical units, bed boundaries and sequences, which can be correlated locally to regionally. Subsurface stratigraphical units have been formally defined based largely on logs, following the procedures outlined under *Lithostratigraphical units*; type wells should be established and results published to allow regional correlation.

The importance of geophysical data generated from boreholes for regional stratigraphy, seismic and sequence stratigraphy, has been particularly well illustrated for the Gulf of Mexico (O'Neill *et al.* 1999) and the North Sea (UKOOA) where they have been instrumental in establishing the regional extent and correlation of subsurface rock units and events (e.g. bed boundaries, unconformities, maximum flooding surfaces, coarsening/fining-up sequences).

3.4. Seismic stratigraphy

Seismic stratigraphy is a method for delineating stratigraphical units from seismic reflection profiles (see Vail *et al.* 1991; Emery & Myers 1996, for reviews and further references). Seismic reflections result from acoustic impedance contrasts and are dependent upon abrupt changes in the density and/or velocity attributes of adjacent strata. Depending on the frequency of the seismic source used, seismic reflections can yield information about stratal arrangement from depths of a few metres to many kilometres. The stratigraphical information derived may be two dimensional or three dimensional. Stratal packages can be defined, based on consideration of three main facets of the seismic data set: (a) reflector termination patterns (b) prominent and continuous reflectors, and (c) seismic 'facies' or areas of the profile showing particular reflection characteristics such as frequency, continuity, amplitude, shape, etc.

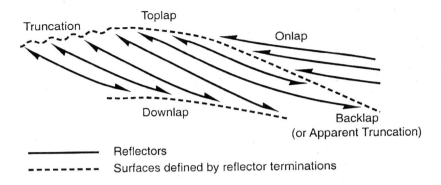

Fig. 3.4. Commonly-used terms to describe patterns of reflector termination in seismic stratigraphy.

Because reflectors represent bedding interfaces, rather than lateral facies changes, they are for practical purposes treated as isochronous surfaces. Exceptions to this general rule are the distinctive reflectors that come from fluid contacts and burial-related diagenetic horizons, or those that are artifacts of the seismic reflection method (e.g. the 'multiples'). Figure 3.4 illustrates the common reflector termination patterns encountered in reflection profiles. Many of these patterns are the same as employed in the definition of depositional sequences (Fig. 3.5), and seismic stratigraphy may be the starting point of a sequence stratigraphical analysis.

Terminology has not been standardized, but there are two main categories of reflector terminations – **lapout**, which is the termination of a reflector at its depositional limit, and **truncation**, where the rocks represented by the reflector have been terminated by erosion, faulting or intrusion. Beyond this basic distinction, reflectors may be further characterized by either down-dip or up-dip termination, where 'dip' refers to the orientation of the original depositional surface (Fig. 3.5). Up-dip terminations are **onlap** (against an underlying surface) and **toplap** or **erosional truncation** (against an overlying surface), the latter two being grouped by some authors as **offlap**. Down-dip terminations are **downlap** against an underlying surface, and **apparent truncation** against an overlying surface. As a better alternative to apparent truncation, the term **backlap** has been used by some authors (Kidwell 1991). In slope settings true down-dip erosional truncation is commonly observed.

3.5. Sequence stratigraphy

Sequence stratigraphy is concerned with the large-scale, three-dimensional arrangement of sedimentary strata, and the major factors that influence their geometries, namely sea-level change, contemporaneous fault movements, basin subsidence and sediment supply

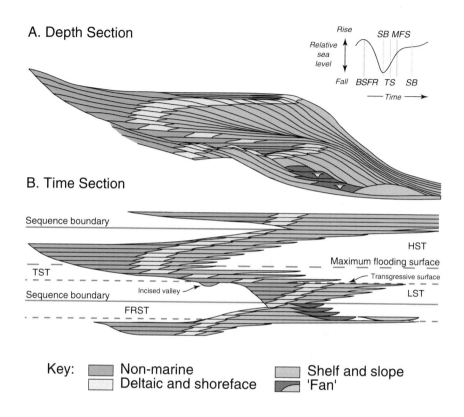

Fig. 3.5. Generalized model of a depositional sequence. In (**a**) the vertical axis is depth; the same sequence is illustrated in (**b**) but with the vertical axis as time. Sediment supply is from the right, and different depositional environments are indicated. No temporal or thickness scale is given because sequences clearly develop in a hierarchical fashion at a great range of scale.

(e.g. reviews by Emery & Myers 1996; Miall 1997). The observational basis of sequence stratigraphy is the ubiquitous arrangement of strata into units bounded above and below by unconformities that can be traced out into conformable surfaces in a basinward direction. These surfaces are defined as the **sequence boundaries** and the strata between them constitute a **depositional sequence**.

The geometrical relationships are observable from seismic reflection profiles, extensive (e.g. hillside) field exposures, or are inferred by correlations from smaller locations. A generalized model of a depositional sequence, including details of the internal geometries, is shown in Fig. 3.5. No temporal or thickness scale is given in this figure because sequences develop in a hierarchical fashion at a great range of scales (e.g. Van

Wagoner *et al.* 1990). The principal factor thought to govern the genesis of a depositional sequence is relative sea-level change. Relative sea-level change is the net result of global sea-level change combined with local subsidence or uplift of the depositional area (Posamentier & Vail 1988).

An important practical aspect of sequence stratigraphy is the recognition of key surfaces. In the Exxon model of sequence stratigraphy (e.g. Haq *et al.* 1988) the **unconformity surface** represents the proximal area of a sequence boundary and passes into expanded sedimentary successions in more basinal settings; it develops when proximal accommodation is no longer available. Accommodation is a loosely defined term meaning space available for sediment to accumulate: this space is capped by a dynamic 'accommodation limit', a surface which passes through the shoreline and is itself dependent on sediment supply and transport processes. The proximal region of the sequence boundary is characterized by erosion of the underlying strata and may include such features as river valleys cut into previously deposited marine strata of the underlying surface. Conversely, the distal region of the sequence boundary may be represented by the increased volume of sedimentary debris eroded from the more landward sites.

In addition to the sequence boundary, two other important surfaces occur within a sequence. The more distal portion of the **maximum flooding surface** represents, like the sequence boundary, a break in deposition or very slow sedimentation, but unlike the sequence boundary it develops at the far end of the sediment transport path as a result of sediment starvation, and may be characterized by condensed marine deposits containing an abundant pelagic fauna and well-developed early (sea-floor) authigenic mineralisation, especially with glauconite and phosphate. The more landward portions of the maximum flooding surface may, by contrast, be hidden within a thick succession of relatively shallow marine or non-marine strata. (It is the maximum flooding surface that normally defines the limits of Galloway's (1989) genetic stratigraphical sequences: see below). A third important surface is the **transgressive surface**, which is generally taken to be the first significant marine flooding surface within the sequence.

Within sequences, further, more subtle geometrical and facies relationships have been used to define **systems tracts** (Van Wagoner *et al.* 1988; cf. Helland-Hansen & Gjelberg 1994; Helland-Hansen & Martinsen 1996). Geometrical arrangements of facies or smaller-scale sedimentary cycles ('parasequences') may be such that systems tracts can be recognized in single vertical sections at outcrop or within a borehole (Van Wagoner *et al.* 1990). A four-systems-tract subdivision of depositional sequences is now commonly employed (Hunt & Tucker 1992, 1995). Overlying the sequence boundary is the **lowstand systems tract**, characterized by inferred rising relative sea-level and shoreline regression (the latter continuous from the preceding systems tract). The **transgressive systems tract** comprises strata whose depositional environments migrate overall in a landward direction (i.e. are transgressive) and whose component stratal surfaces onlap pre-existing deposits; the base is defined by the transgressive surface and relative sea-level at the shoreline is also inferred to have been rising. The transgressive systems tract is terminated at its top at the maximum flooding surface, above which strata of the **highstand systems tract** shift basinward again, with successive stratal surfaces terminating in progressively more distal locations, forming a geometrical pattern known as downlap (note the general similarities with the lowstand systems tract). The final, **forced regressive**

systems tract is represented by an arrangement of strata whose shoreline positions migrated progressively downwards as well as basinwards and so is produced during falling relative sea-level and regression (the surface defining the base had been termed the 'basal surface of forced regression').

The timing and nature of gravity-flow deposits within a depositional sequence is a matter of some discussion. Early models (e.g. Posamentier & Vail 1988) placed major phases of debris flow and turbidity current deposition as occurring during relative sea-level fall and lowstand. More recent work (e.g. Helland-Hansen & Gjelberg 1994) has shown that, conceptually, gravity-driven deposition may occur at any stage of relative sea-level change, depending on the development of steep and unstable slopes. In Figure 3.5 'fan' systems are shown in the more conventional position corresponding to relative sea-level fall.

Carbonate sediments respond in a different manner to siliciclastic sediments in response to relative sea-level change because the sedimentary grains are produced *in situ* rather than transported from a hinterland. Many carbonate-producing processes require warmth and light in shallow-water settings and are particularly sensistive to changes in nutrient supply. Thus the large-scale geometries of carbonate systems are very different from those of siliciclastic systems (Schlager 1992). During relative sea-level rise carbonate production may be extremely effective, such that large volumes of shallow water carbonate accumulate, and there is sufficient production for export of carbonate grains to deep water settings. During relative sea-level falls, large regions of potentially productive shelf may be exposed and effectively shut down, so that little sediment can accumulate in either shallow-water or deep-water settings. However, it is probable that in the past some sea-level rises have been associated with increases in nutrient supply and/or siliciclastic sediment that have led to the demise of the carbonate platform systems, to form a **drowning unconformity** (Schlager 1989). The lower depositional angles of the siliciclastic sediments may create stratal geometries that mimic those expected when sea-level falls below the shelf-slope break of purely siliciclastic systems.

It has been claimed by some that global sea-level change is the dominant influence on the formation of all depositional sequences (e.g. Haq *et al.* 1988), but this is not accepted as fact by the majority of workers in the field. The extent to which global sea-level change may influence the sequence stratigraphical record will depend on both time and location. In many depositional settings (for example, active plate margins) more localized relative sea-level changes commonly dominate the stratigraphical record. On the other hand at certain times, e.g. during periods of glaciation and deglaciation, high magnitude, high rate sea-level fluctuations have undoubtedly occurred.

Sequence stratigraphy is now widely used as a means of subdividing, correlating and dating sediments (e.g. Hesselbo & Parkinson 1996), especially within the hydrocarbons industry. But in many successions, for example in the Upper Jurassic to Lower Cretaceous sequences of the North Sea Basin, Galloway's (1989) genetic sequences are the type recognized, because a maximum flooding surface or condensed horizon is easily detected in geophysical log data (Rawson & Riley 1982; Partington *et al.*, 1993a, b).

It should also be noted that sequence stratigraphy is often applied uncritically; in particular, age assignment of strata based solely on correlation with a supposedly global sea-level curve has not proved to be a robust method.

3.6. Cycles in stratigraphy

Transgressive–regressive cycles occur at a variety of scales in the geological record and have long been utilized in both division and correlation of stratigraphical sequences. In his study of the UK Namurian sequences, Ramsbottom (1977) distinguished cycles of different scales, from **synthem** through **mesothem** down to the small-scale **cyclothem**. Cyclothem is the one that is most used and is scale independent.

Over the ten years since the last edition of this guide, there has been a huge surge of interest in the study of mesothem and cyclothem scale sedimentary cycles that have been generated by changes in the Earth's orbit. The investigation of the frequently complex way in which orbital cycles have influenced Earth's climates, oceans, and ice-caps, and of the resulting cycles in the stratigraphical record, is called **cyclostratigraphy** – a term first used at a meeting in Perugia, Italy, in 1988.

A Serbian mathematician, Milutin Milankovitch, calculated variations in Earth's orbit (of tens to hundreds of thousands of years in duration) accurately for the first time, and they are now called **Milankovitch Cycles**. Milankovitch (1941) showed quantitatively how these orbital cycles determined the amount of solar radiation reaching the Earth, and proposed that they were responsible for climatic changes leading to the ice ages. His work was not treated seriously at the time because the changes in solar energy he calculated seemed much too small to have caused glacial and interglacial periods. Milankovitch cycles remained unproven in the stratigraphical record until Hays *et al.* (1976) demonstrated regular cyclicity of 100 kyr, 41 kyr, 23 kyr and 19 kyr in oxygen isotope and other data in deep sea cores covering the last 500 kyr. The periodicity exactly matched major orbital cycle frequencies previously calculated by Milankovitch, and demonstrated that orbital cycles acted as a pacemaker to the ice-ages. Furthermore, the small changes in solar energy that previously had caused scientists to doubt Milankovitch's theory were shown to be strongly amplified by positive feedback mechanisms.

Identification of sedimentary cycles whose deposition was controlled by Milankovitch Cycles has enabled geologists to develop an orbital time-scale graduated in tens to hundreds of thousands of years which now extends back 20 myr to the base of the Miocene.

Identification of orbital cycle frequencies in older sediments has not enjoyed such dramatic success. Although regular bedding cycles which fall in the Milankovitch frequencies are conspicuous features of many Phanerozoic sedimentary successions, precise estimates of cycle frequency has often proved difficult to achieve. This is probably because of the numerous other time-dependent variables which affect sedimentation, such as rapidly changing accumulation rate and hidden hiatuses. Nevertheless, studies such as those in House & Gale (1995) and Shackleton *et al.* (1999) have shown the potential for using cyclostratigraphy for creating an orbital time-scale for part of the Mesozoic and for independently calibrating the duration of ammonite biozones. The effects of Milankovitch climate forcing have also been recognized in some non-marine sequences, such as in the lacustrine Triassic sediments of the Newark Basin, USA (Olsen & Kent 1996).

3.7. Climate stratigraphy

For the Quaternary there has been a tradition of dividing sediment sequences on the basis of climatic changes, particularly the glacial deposits of central Europe and mid-latitude North America. This approach has remained central to the subdivision of Quaternary successions. The recognition of climatic events from sediments is not straightforward on land, on the shelves, or in the deep ocean (where the modern isotope stage sequence has been defined).

In the past, the preferred climatic time-scale was that developed by Penck & Brückner (1909–11) for the Alps. In recent decades it has been replaced by the oxygen isotope record (see *Isotope Stratigraphy* p. 31). Today, the burden of correlation lies in equating highly fragmentary, yet high resolution, terrestrial and shallow-marine sediments on the one hand, with the potentially continuous, yet comparatively lower resolution ocean isotope sequence on the other.

The commonly used geological-climate divisions 'glacial' and 'interglacial' are very difficult to apply outside glaciated regions, i.e. most of the world, and many of the 'glacials' were simply cold periods. Hence, Suggate & West (1969) proposed the terms **cold** (= glacial) and **temperate** (= interglacial) stages instead. There remains a problem of where to draw the boundaries between them. Where possible, as in ocean sediment cores, boundaries are generally placed at mid-points between temperature maxima and minima.

The original intention was that cold or warm (= temperate) stages should represent the first-rank climatic oscillations recognized, and that subdivision of these would be into **substages** or **zones**. However, high-resolution investigations have allowed the recognition of ever more climatic oscillations of decreasing intensity or wavelength within the first-rank divisions. These events are stretching the ability of the stratigraphical terminology to cope with the escalating number of names they generate. Terms such as **event**, **oscillation** and **phase** are currently in use to refer to short or small-scale climatic events. Clear patterns of hierarchy are becoming blurred. As our ability to resolve smaller and smaller oscillations increases, a more detailed nomenclature will inevitably emerge.

For many Quaternary workers climato- and chronostratigraphical terminology are interchangeable. Although this situation is clearly unsatisfactory, because of the imprecision that it may bring to inter-regional and ultimately global correlation, it is likely to continue for the foreseeable future. The long-term goal should be to develop a formally defined, chronostratigraphically based system that is fully compatible with that for the rest of the geological column, supported by reliable geochronology. This may be some way off, however; for example, in Europe some of the chronostratigraphical terms in current use do not have defined boundary or unit stratotypes. Moreover, because climatic change is used as the basis for chronostratigraphy and because climate change is represented differently under terrestrial and marine conditions, it is necessary that terrestrial and marine successions are clearly separated (cf. Gibbard & West 2000). Reliance on a single parameter-based, global scheme, such as the marine oxygen isotope sequence (e.g. Bowen *et al.* 1999), raises many practical difficulties and is the subject of much current research.

Steps are being taken to develop a chronostratigraphically-based system through the International Quaternary Association's (INQUA) Subcommission on European

Quaternary Stratigraphy (Turner 1998). Unfortunately, some Quaternary workers do not see the need for this, especially those who rely on geochronology, e.g. radiocarbon, for correlation. For example, despite repeated attempts to propose a new GSSP boundary stratotype for the Pleistocene–Holocene boundary (Olausson 1982), none has yet been universally accepted.

3.8. Event stratigraphy

Event stratigraphy comprises the study of stratigraphical traces of relatively short-lived events (instant to thousands of years) compared to those normally observed on a geological time-scale. Events may be represented by depositional, erosional, or geochemical features. They may be of local significance (e.g. a debris flow), or more extensive (e.g. a volcanic ash deposit), or even global (eustatic flooding surface). They may be random (discyclic) or regular (cyclic).

Virtually instantaneous events (e.g. rainprints, footprints) are mostly of local significance, but a bolide impact may produce tektite bands or isotopic excursions in sediments covering wide areas, e.g. at the end of the Cretaceous in Central and North America. Other short-term events may record periods lasting from minutes to a few days, such as tempestites, some turbidites (gravitational flow deposits), tsunamiites (tidal wave deposits), fusain layers (e.g. in the Lower Carboniferous of Donegal), lava flows and flood deposits.

Medium-scale events (days to several years) include glacial varves and some flood deposits. Palaeobiological events include dinoflagellate or nannofossil blooms, which provide regional markers and may cause anoxic events. Mass mortality may be represented by unusually fossiliferous horizons (e.g. some lagerstätten).

Long-term events (tens to several thousand years or more) are represented by palaeosols, transgressive flooding surfaces, hard grounds, omission surfaces, septarian nodule horizons, palaeomagnetic polarity reversals (about 3–4 kyr duration), most evaporite deposits and diamictites. Palaeosols may record a complex history of fluctuating water tables, due to climatic or base level events. Evolutionary or migratory appearances of organisms form the basis for biostratigraphical event correlation (see *Biostratigraphy* p. 15).

Collating all the event stratigraphy within and between basins can provide a stratigraphical resolution which normally exceeds that of biostratigraphy. A good example is the interlayering of K bentonites (ashes from Plinian volcanoes, Spears *et al.* 1999) and glacioeustatic flooding events (marine bands) in parts of the Namurian (Carboniferous) of Europe, which provides an average resolution of less than 65,000 kyr duration between the base of each flooding surface.

3.9. Magnetostratigraphy

Magnetostratigraphy exploits variations in the magnetic properties of rocks as a basis for geological correlation (e.g. Hailwood 1989; Opdyke & Channell 1996). The most widely used property is the direction of the primary remanent magnetism, which records the geomagnetic field polarity at the time of formation of the rock. Sets of magnetic polarity reversals in sedimentary sequences can be correlated between sections. Furthermore, biostratigraphical information from the sequences often permits correlation of the observed polarity sequence with the appropriate part of the radiometrically-calibrated geomagnetic polarity time scale (GPTS). This allows the assignment of a numerical age to each of the beds containing a polarity reversal within the succession. Because polarity reversals effectively happen simultaneously over the whole surface of the earth, they can be used for global correlations.

The pattern of geomagnetic polarity reversals that forms the basis of the GPTS for the Cenozoic and late Mesozoic is well established from ocean floor spreading magnetic anomalies, so that the GPTS for these periods is now well-defined (e.g. Berggren *et al* 1995). However, the general absence of early Mesozoic and older sea floor, together with the corresponding magnetic anomaly record (due to its removal by subduction) leads to increasing uncertainty in the GPTS for older periods. A further problem in magnetic stratigraphy is the occurrence of magnetic overprints, acquired during later thermal or chemical events in the rock's history. Often, these can be removed by incremental demagnetisation, to isolate the primary component. However, in some circumstances the overprint is more stable than the primary component, or it may completely replace the latter.

For the late Mesozoic to Recent the GPTS is age-calibrated by combined palaeomagnetic and radiometric age investigations of terrestrial lava sequences such as those on Iceland, and by Milankovitch cyclicity in sedimentary sequences. For earlier periods, this information is supplemented by radiometric age determinations on glauconites and U–Pb dates on zircons in interbedded tuffs in sedimentary sequences which contain polarity reversal records. A further source of information is biostratigraphical age determinations on sediments which immediately overly specific ocean floor spreading magnetic anomalies in the ocean basins and which have been cored by deep sea drilling.

The basic magnetostratigraphical unit is the magnetic zone or **magnetozone**, which usually is identified as a rock interval characterised by a specific (either normal or reverse) dominant magnetic polarity. The interval of time corresponding to a particular magnetic polarity zone is the **magnetic chron,** which has a typical duration of about 10^5 to 10^7 years. **Subchrons** are shorter intervals of opposing polarity within a chron and **superchrons** are longer intervals of dominantly normal, reverse or mixed polarity. Small scale perturbations in ocean floor magnetic polarity records which may represent very short geomagnetic polarity events are called **cryptochrons.**

The four most recent magnetic chrons, which extend from the Recent to the late Miocene, were named after pioneering workers in geomagnetism. In younger to older order, these are the Brunhes normal, Matuyama reverse, Gauss normal and Gilbert reverse polarity chrons. Each of these contains distinctive subchrons, such as the Olduvai subchron which is located within the Matuyama chron, close to the Pliocene–Pleistocene boundary. These and earlier chrons are now conventionally labelled from the corresponding ocean-floor spreading magnetic anomaly number. The number is usually suffixed by the letter *n* or *r*, according to whether the dominant magnetic polarity is normal or reverse. It is prefixed by the letter C (Cenozoic) to discriminate the time interval (e.g. chron C21n) from the corresponding magnetic anomaly number (e.g.

anomaly 21).

A magnetic zone may sometimes be defined in terms of magnetic susceptibility or some other distinctive magnetic property of the rock, instead of the magnetic polarity. Magnetic susceptibility is essentially a mineralogically-controlled parameter, that reflects the composition, concentration and grain size of magnetic minerals within the rock. Fluctuations in susceptibility may reflect climatic, tectonic or other controls on sedimentation, that can be correlated locally, regionally or globally. For example, susceptibility fluctuations in loess sequences and deep marine successions have been correlated with oxygen and carbon isotope stages, reflecting global climatic changes.

The magnetostratigraphical record is being applied increasingly to help resolve correlation problems. An interesting example is in the study of Milankovitch climate forcing in the Newark basin (see *Cycles in stratigraphy* p. 27), where the polarity record provided an independent check on correlation of the individual core sections that build up the composite log sequence (Olsen & Kent 1996).

3.10. Isotope stratigraphy

Isotope stratigraphy is a method of determining relative ages of sediments based on measurement of isotopic ratios of a particular element. It works on the principle that the proportions of some isotopes incorporated in biogenic minerals (calcite, aragonite, phosphate) change through time in response to fluctuating palaeoenvironmental and geological conditions. However, this primary signal is often masked by diagenetic alteration of sediments which have secondarily altered the isotopic ratios. Disentangling primary and secondary components of measured isotopic ratios is a difficult and frequently controversial subject. Although isotopes of many elements have been studied, strontium, oxygen and carbon are of particularly wide application (Marshall 1992; McArthur 1998).

3.10.1. Strontium isotopes

Strontium isotope stratigraphy relies on measurement of $^{87}Sr/^{86}Sr$ in marine biogenic carbonate or phosphate. Precipitation of these minerals involves incorporation of strontium from seawater, which will have an $^{87}Sr/^{86}Sr$ identical to that of oceanic values, which is of the same value globally at any point in time. The $^{87}Sr/^{86}Sr$ ratio changed systematically through time and it is therefore possible to date samples by placing them on a standard curve. The method works best for periods of time over which there was a long-term unidirectional shift in ratios, as during the Tertiary. Strontium isotope stratigraphy gives a maximum time-resolution of about 1 myr.

Strontium in seawater is derived from three sources: fluvial input of material weathered from continental crust; hydrothermal leaching of oceanic basalts at mid-ocean ridges; and recrystallization of carbonate minerals. Changing strontium values reflect global changes in these geological processes.

3.10.2. Oxygen isotopes

The ratios in which the two stable isotopes of oxygen (^{16}O and ^{18}O) are precipitated in carbonates and phosphates depends upon the oxygen isotopic composition of the fluid from which the mineral precipitated and also on the temperature at which this took place. However, some organisms incorporate oxygen isotopes that are out of equilibrium with temperature and seawater composition. In addition, primary isotopic values may commonly be altered by diagenetic recrystallisation of carbonate sediments.

Oxygen isotopes can record detailed changes in ocean temperature and ice volume. The most extensive use of oxygen isotopes has been in deep sea cores of Cenozoic and Quaternary sediments, where data from calcitic microfossils, notably foraminifera, record fluctuating temperatures and the growth and decay of ice-sheets, allowing the recognition of oxygen isotope stages. The separate effects of temperature and ice volume are distinguished by comparing isotope ratios in coeval planktonic and benthonic microfossils, mainly foraminifera. Because both parameters were driven by Milankovitch climatic cycles, it has been possible to identify and correlate oxygen isotope stages in detail across the globe, and ^{18}O curves provide a very refined (20 kyr resolution) timescale for Neogene time. In pre-Cenozoic sediments the use of oxygen isotopes in both stratigraphy and palaeoenvironmental studies has been much more limited because much of the carbonate is recrystallized, and only rarely reflects secular changes in oxygen isotope ratios.

3.10.3. Carbon isotopes

The two stable isotopes of carbon, ^{12}C and ^{13}C, vary in relative abundance through time in both carbonate minerals and organic matter. The fluctuations in ^{13}C are brought about by changes in the balance of fluxes of the carbon cycle, including inputs of terrestrial carbon and oxidation of marine organic matter, and outputs by production and burial of marine carbonate and organic matter. Because the residence time in the carbon cycle is brief (10 kyr), changes in flux are recorded accurately and globally in the sedimentary record. Furthermore, carbon isotopes are relatively robust and resistant to diagenesis.

In the Mesozoic, large positive excursions in ^{13}C have been encountered in the Toarcian, the Aptian and the Cenomanian. These are globally synchronous, and are interpreted as a consequence of sequestration of ^{12}C by the burial of organic matter as black shales in the ocean basins and outer shelves. They provide a global, high-resolution means of correlation.

3.11. Chemostratigraphy

There can be a special need to attempt correlation in biostratigraphically barren rock sequences occurring, for example, in hydrocarbon-rich basins. Such cases are often scale-dependent and within the normal resolution of biostratigraphical practice: for example, within an oil field reservoir where fine correlation is required in sandstones that lack fossils. In such circumstances a variety of chemostratigraphical methods can be employed with varying degrees of success. These range from the recognition of distinctive heavy mineral suites (Mange-Rajetsky 1995) to chemical analysis of vertical stratigraphical profiles (Preston *et al.* 1998). Such chemical analyses are relatively rapid, straightforward and require very small samples following the advent of ICP based systems. Examples of correlation using chemical methods include Wray & Gale (1993, marl bands in the Chalk) and Pearce & Jarvis (1995, Quaternary turbidites), who analysed specific marker beds and showed high resolution correlations to the level of a single sample horizon.

Correlation can also be achieved using stratigraphical trends in clay mineral assemblages. Such trends usually reflect long-term changes in sea level, climate or tectonic regime. More rarely, they reflect relatively short-term events, in which case they provide a means of relatively precise correlation. Jeans (1995) discusses the application of clay mineral stratigraphy to red-bed successions in the Palaeozoic and Mesozoic of the UK.

Events can be recorded geochemically in, for example, diagenetic cement histories (as observed by cathodoluminescence-cement stratigraphy, e.g. Lees & Miller 1985), early hydrocarbon impregnation, the appearance of minerals and trace elements reflecting provenance change, or as isotopic events, either in sediments or in biominerals.

4. Establishing a relative time-scale: chronostratigraphy

Chronostratigraphy ('time stratigraphy') is that branch of stratigraphy concerned with the application of time to rock successions, leading ultimately to the establishment of a globally applicable standard time-scale. The chronostratigraphical scale (Figs 4.1–4.3) originated as a relative one which has been built up piecemeal over the last 200 years, primarily through the application of biostratigraphy. Reliable numerical ages have been added gradually to form a parallel, numerical time-scale based on years.

The chronostratigraphical scale and the corresponding divisions of geological time are:

Chronostratigraphical divisions	Geochronological divisions
eonothem	eon
erathem	era
system	period
series	epoch
stage	age
chronozone	chron

Chronostratigraphical divisions are 'time/rock' units, i.e. they refer to the sequence of rocks deposited during a particular interval of time. Geochronological divisions are the corresponding (abstract) intervals of geological time. Thus, we can say that rocks of the Cretaceous *System* were deposited during an interval of time called the Cretaceous *Period*. This distinction dates back to the early days of our science, but the present-day need for it has been challenged (e.g. Harland *et al.* 1990) and is currently being reviewed by the Stratigraphy Commission.

Many of our chronostratigraphical units were originally defined rather loosely. Hence modern work focuses on the rigorous definition of each component of the time-scale. For the Phanerozoic, it is based on the principle that the base of each chronostratigraphical division should be defined at a specific level in a type section, the **Global Stratotype Section and Point** (GSSP). The top of a division is automatically defined by the base of the overlying division. The base of a system equates with the lower boundary of the appropriate series or stage. Procedures for defining a GSSP are described by Remane *et al.* (1996). A 'golden spike' is placed at the GSSP, marking the unique place where a specified point in time is indicated. Sections elsewhere are then correlated with this using all possible methods. Hence, the GSSP must be at a level and locality that has maximum possible applicability for global correlation, although in practice no single GSSP is likely to be directly correlatable over the whole world, so that supplementary reference points may be necessary.

The IUGS's International Commission on Stratigraphy, through its various subcommissions and working groups, is responsible for the international collaboration necessary to achieve global agreement on the choice of individual GSSPs. Once formal

SYSTEM	SERIES	STAGE	GSSP
Quaternary	Holocene		
	Pleistocene		1985
Neogene	Pliocene	Gelasian	1996
		Piacenzian	1997
		Zanclean	1999
	Miocene	Messinian	1999
		Tortonian	
		Serravallian	
		Langhian	
		Burdigalian	
		Aquitanian	1996
Paleogene	Oligocene	Chattian	
		Rupelian	1993
	Eocene	Priabonian	
		Bartonian	
		Lutetian	
		Ypresian	
	Paleocene	Thanetian	
		Selandian	
		Danian	1991

Fig. 4.1. The Cenozoic chronostratigraphical scale, showing agreed GSSPs and the years in which they were ratified.

proposals are made by the appropriate subcommission they are voted upon by members of the International Commission on Stratigraphy. The results are then published in *Episodes*, the official journal of the IUGS. So far only a few system, series and stage boundaries have been agreed; these are indicated on Figures 4.1–4.3.

Where an existing chronostratigraphical name is formalized, historical priority may be considered in defining a GSSP, but will not take precedence over the needs for the GSSP to provide the best possible potential for correlation. Thus, although the Devonian System was named after Devon (SW England), the base of the system is not seen there and the basal GSSP is now defined at the level where *Monograptus uniformis* was observed to appear at Klonk in the Czech Republic. It should be noted that the level of the GSSP would not change if further study should show that *M. uniformis* appears lower down in the section.

SYSTEM	SERIES	STAGE	GSSP
Cretaceous	Upper	Maastrichtian	2001
		Campanian	
		Santonian	
		Coniacian	
		Turonian	
		Cenomanian	
	Lower	Albian	
		Aptian	
		Barremian	
		Hauterivian	
		Valanginian	
		Berriasian	
Jurassic	Upper (Malm)	Tithonian	
		Kimmeridgian	
		Oxfordian	
	Middle (Dogger)	Callovian	
		Bathonian	
		Bajocian	1996
		Aalenian	2000
	Lower (Lias)	Toarcian	
		Pliensbachian	
		Sinemurian	2000
		Hettangian	
Triassic	Upper	Rhaetian	
		Norian	
		Carnian	
	Middle	Ladinian	
		Anisian	
	Lower	Olenekian	
		Induan	2001

Fig. 4.2. The Mesozoic chronostratigraphical scale, showing agreed GSSPs and the years in which they were ratified.

The definition of the earlier Precambrian chronostratigraphical divisions follows a different route. Instead, geochronology is applied; for example, the Proterozoic–Archaean boundary is commonly defined as being at 2500 Ma. This dichotomy raises problems of definition. In particular, the top and bottom of the Proterozoic eon will be defined in different ways; the top as a GSSP marking the base of the Cambrian, and the base (probably) as at 2500 Ma.

SYSTEM	S-S	SERIES	STAGE	GSSP
Permian		Lopingian	Changhsingian	
			Wuchiapingian	
		Guadalupian	Capitanian	2001
			Wordian	2001
			Roadian	2001
		Cisuralian	Kungurian	
			Artinskian	
			Sakmarian	
			Asselian	1996
Carboniferous	Upper	Gzhelian		
		Kasimovian		
		Moscovian		
		Bashkirian		1996
	Lower	Serpukhovian		
		Visean		
		Tournaisian		1991
Devonian		Upper	Fammenian	1993
			Frasnian	1991
		Middle	Givetian	1995
			Eifelian	1985
		Lower	Emsian	1996
			Pragian	1989
			Lochkovian	1977
Silurian		Prídolí		1985
		Ludlow	Ludfordian	1985
			Gorstian	1985
		Wenlock	Homerian	1985
			Sheinwoodian	1985
		Llandovery	Telychian	1985
			Aeronian	1985
			Rhuddanian	1985
Ordovician		Upper		
		Middle	Dariwillian	1997
		Lower		2000
Cambrian		Upper		
		Middle		
		Lower		1994

Fig. 4.3. The Palaeozoic chronostratigraphical scale, showing agreed GSSPs and the years in which they were ratified.

4.1. Chronostratigraphical terminology in practice

- At any level in the chronostratigraphical hierarchy, an initial capital letter is used for each formal component of the name: e.g. Jurassic System, Hettangian Stage.
- Chronostratigraphical units may be divided **formally** into Lower and Upper, or Lower, Middle and Upper, the corresponding geochronological units into Early and Late, or Early, Mid and Late. Use of lower case initial letters (lower, mid, etc.) implies an **informal** usage, either because it is being used more loosely or because a unit may not have been formally divided yet.
  Note our recommended usage of 'Mid' for geochronological units, which follows the last edition of this guide. North American authors normally use 'Middle' (following Hedberg 1976), which thus fails to distinguish geochronological units from chronostratigraphical ones.
- 'Series' should be used only as a chronostratigraphical division and not as a lithological term, the 'Atherfield Clay Series' is now the Atherfield Clay Formation. Series names should not normally end in -ian, e.g. 'Wenlock Series' not 'Wenlockian Series', but note that Carboniferous series names do end thus.
- Stage names should preferably be based on a geographical name and end in -ian, e.g. 'Hauterivian'. A stage is the smallest chronostratigraphical unit that normally can be recognized globally, and typically has a duration of some 5–10 myr.

5. Determining a numerical time-scale: geochronometry

Geochronometry is the measurement of geological time to produce a numerical time-scale (not 'absolute', as there is always a margin of error). It applies geochronological methods, especially radiometric dating. The geochronological scale is a periodic scale using the year as a basic unit. Apparent ages obtained in geochronometry are referred to as radiometric or isotope dates. For older rocks, multiple annual units are normally written in thousands of years (ka), million years (Ma) or, in the case of the oldest rocks, in thousands of millions of years (Ga); Holocene and Pleistocene dates are normally quoted in years before 1950 (years BP (before present)). Note that although the **duration** of an interval is normally expressed differently from its **age**, there is no international 'standard': we recommend **kyr** and **myr**.

Rank terms of geological time (eon, era, period, epoch and age) may be used for geochronometrical units when such terms are formalized. For example, the Archaean Eon and the Proterozoic Eon could be formal geochronometrical units distinguished on the basis of an arbitrarily chosen boundary at 2.5 Ga (2500 Ma).

Decay schemes that can be used for geochronology have to fulfil several criteria; they have to have an isotope with a long enough half life to be useful over the period of geological time and the half life has to be known accurately. In addition, the element has to exist in sufficient quantity in the rocks and minerals under study to be extracted and analysed. There are now many different isotope decay schemes in use for geochronological purposes and, because of varying chemical and mineral stability during geological events, complex geological histories can be deduced by targeting problems with a suitable geochronometer. For instance, in the case of a granite that was deformed and developed a schistosity and was later cut by localized shears, it should be possible to date granite emplacement from U–Pb on zircons, use Rb–Sr on mica and feldspar to date the schistosity and use K–Ar on mica in the shear zone to look at the time of reheating during localized shearing. It is important to know what event or process is under scrutiny and then to choose an appropriate geochronological tool. Good descriptions of techniques and their applications can be found in Faure (1986), Roth & Poty (1989), Dickin (1995) and Richards & Noble (1998).

The two techniques most commonly used by stratigraphers today are $^{40}\text{Ar}/^{39}\text{Ar}$ of potassium-bearing minerals and U–Pb analysis of zircons, because both these methods can provide high precision ages. The $^{40}\text{Ar}/^{39}\text{Ar}$ method is based on the decay of potassium to the inert gas argon, which becomes physically trapped in the crystal lattice on formation. A reliable age is dependent upon the argon being held in place in substantial parts of the crystal. The commonly used step heating method, which involves progressive degassing of the samples up to melting point and analysis of the argon from each step, provides a way of looking at argon loss from different parts of the lattice and enables well-preserved parts of the crystal yielding crystallization ages to be distinguished from those which have suffered argon loss.

U–(Th)–Pb analysis of zircon is the main tool for determining the crystallization age of igneous rocks. The robust nature of zircon means that its crystallization age can be retained though major disturbances, and the integrity of its isotope composition can be checked because the method uses two decay schemes; ^{235}U and ^{238}U. When these two schemes give the same result and the data plot on the concordia (the curve defined by the undisturbed accumulation of radiogenic Pb from ^{235}U and ^{238}U) the mineral is said to be concordant and recording the time of crystallization.

The main application of geochronology in stratigraphy is the calibration of the time-scale. This requires the combination of well-defined stratigraphical units interbedded with material suitable for radiometric dating. Volcanic ashes and their altered bentonite equivalents represent short-lived eruptions. They are laterally extensive and cross facies boundaries, thus providing excellent time planes within the stratigraphical record. Although the original volcanic glass has usually been converted to clay, crystalline igneous minerals are commonly preserved. U–Pb dating of zircons and $^{40}Ar/^{39}Ar$ dating of micas and sanidine from such deposits have provided some of the most precise calibration of the time-scale in recent years (Huff *et al.* 1997; Kowallis *et al.* 1995; Bowring *et al.* 1998; Tucker *et al.* 1998).

6. Combining the approaches: holostratigraphy

Holostratigraphy is a holistic approach to stratigraphy – it brings together every possible method to produce an integrated correlation that may have a much higher resolution than any one method alone can provide.

In practice, successful holostratigraphy requires a team approach to focus on a particular problem, such as the subdivision and correlation of key sequences or the definition of a chronostratigraphical boundary. For a single stratigraphical section one may, for example, calibrate against a detailed lithological log several different biostratigraphical schemes from different taxonomic groups, together with isotope curves, interpreted Milankovitch cycles, and a magnetostratigraphical record. This may provide an intermeshed sequence of 'events' at very closely-spaced intervals. Such an approach is very important when system or stage boundaries are being formally defined by the various IUGS stratigraphical subcommissions, as it may highlight several key events, both biological and non-biological, that individually may not be globally distributed but taken collectively provide a boundary that can be recognized worldwide.

The terms **high resolution** and **ultra-high resolution stratigraphy** are sometimes used as alternative names to holostratigraphy, but both have been used generally in more restricted ways, in particular to embrace what is here described as cyclostratigraphy, or to refer to a particularly detailed approach to integrated biostratigraphy (especially in the oil industry). Indeed, they have also been misused to imply a higher degree of correlation than has really been achieved. We suggest that the lack of precision makes them redundant as formal terms.

7. Databases

For UK users of this guide there are now several important available stratigraphical databases, both hard-copy and electronic. The British Geological Survey (BGS) has compiled a lithostratigraphical database which lists published lithostratigraphical names for onshore and offshore units, indicating which names the BGS currently retains and which it regards as obsolete. This is available to any interested user via the BGS Website (http://www.bgs.ac.uk) and is being updated continuously. The updating is partly based on the work of several BGS committees, each of which is revising the lithostratigraphical nomenclature of a particular area or part of the column.

In an attempt to encourage the wider geological community to use a common stratigraphical nomenclature, the recommendations of individual BGS stratigraphical framework committees are discussed by and refereed through the Geological Society's Stratigraphy Commission prior to endorsement by the Commission. Endorsed reports are available on the BGS Website, and in hard copy as BGS internal reports. At the time of writing three reports have been endorsed. The Stratigraphy Commission and the BGS are also collaborating in an ambitious *Holostratigraphy* programme (Allen & Rawson 1998), again to be published on the BGS Website. A refereed holostratigraphy of the type Ludlow Series (Silurian) is already available.

The correlation of UK sequences is published system-by-system through the Stratigraphy Commission in the *Special Report Series* of the Geological Society, as are a series of reports on various techniques in stratigraphy. Revised versions are published as the originals become outdated. For the offshore area, the BGS and the United Kingdom Offshore Operators' Association (UKOOA) have jointly published a series of volumes on the *Lithostratigraphical Nomenclature of the UK North Sea*, which define offshore units and review their correlation.

Authors of this Handbook

P. F. RAWSON, Department of Geological Sciences, University College London, Gower Street, London WC1E 6BT

P. M. ALLEN, British Geological Survey, Keyworth, Nottingham, NG12 5GG

P. J. BRENCHLEY, Department of Earth Sciences, University of Liverpool, PO Box 147, Liverpool L69 3BX

J. C. W. COPE, Department of Earth Sciences, University of Wales, College of Cardiff, Main Building, Museum Avenue, PO Box 914, Cardiff CF1 3YE

A. S. GALE, Department of Palaeontology, The Natural History Museum, Cromwell Road, London SW7 5BD and School of Environmental Science, University of Greenwich, Medway Campus, Chatham Maritime, Kent ME4 4TB

J. A. EVANS, NERC Isotope Geosciences Laboratory, Kingsley Dunham Centre, Keyworth, Nottingham NG12 5GG

P. L. GIBBARD, Godwin Institute of Quaternary Research, Department of Geography, University of Cambridge, Downing Place, Cambridge CB2 3EN

F. J. GREGORY, Department of Palaeontology, The Natural History Museum, Cromwell Road, London SW7 5BD and Kronos Consultants, 33 Royston Road, St Albans, Herts AL1 5NF

E. A. HAILWOOD, Core Magnetics, The Green, Sedbergh, LA10 5JS

S. P. HESSELBO, Department of Earth Sciences, University of Oxford, Parks Road, Oxford OX1 3PR

R. W. O'B KNOX, British Geological Survey, Keyworth, Nottingham NG12 5GG

J. E. A. MARSHALL, Department of Geology, University of Southampton, Highfield, Southampton SO9 5NH

M. OATES, BG Group, 100 Thames Valley Park Drive, Reading RG6 1PT

N. J. RILEY, British Geological Survey, Keyworth, Nottingham NG12 5GG

A. G. SMITH, Department of Earth Sciences, University of Cambridge, Downing Street, Cambridge CB2 3EQ

N. TREWIN, Department of Geology and Petroleum Geology, University of Aberdeen, Meston Building, King's College, Aberdeen AB24 3UE

J. A. ZALASIEWICZ, Department of Geology, University of Leicester, University Road, Leicester LE1 7RH.

The Stratigraphy Commission

The Stratigraphy Commission of the Geological Society aims to promote vigorously the fundamental role that stratigraphy plays at the core of our science. It has had a long tradition of initiating publications and meetings, resulting in the highly successful Special Report series, several Special Publications, the *Atlas of Palaeogeography and Lithofacies*, and a regularly updated guide to Stratigraphical Procedure, of which this is the latest, expanded version.

Membership of the Commission is essentially on an individual basis and is drawn from industry, academia and the British Geological Survey. It collaborates with the BGS in several areas, most notably in the development of the Web-based HOLOSTRAT project and in refereeing and endorsing the Survey's framework reports, which aim to rationalize British lithostratigraphical nomenclature. The Commission also liaises with the International Commission on Stratigraphy (ICS) and other international bodies, especially in the relation of UK stratigraphy to global classifications and international stratotypes.

References

AGER, D. V. 1993. *The Stratigraphical Record*. 3rd Edition. Wiley, Chichester.

ALLEN, P. M. & RAWSON, P. F. 1998. HOLOSTRAT – A Web-based Talk Shop for Stratigraphers. *Geoscientist*, **8**(11), 14.

ARMENTROUT, J. M. & CLEMENT, J. F. 1991. Biostratigraphic calibration of depositional cycles: a case study in High Island, offshore Texas. *In*: ARMENTROUT, J. M. & PERKINS, B. F. (eds) *Sequence Stratigraphy as an Exploration Tool. Concepts and Procedures*. 11th Annual Conference, Society of Economic Paleontologists and Mineralogists, 21–51.

BANNER, F. T. & BLOW, W. H. 1965. Progress in the planktonic foraminiferal biostratigraphy of the Neogene. *Nature*, **208**, 1164–1166.

BERGGREN, W. A. & VAN COUVERING, J. A. 1974. The Late Neogene biostratigraphy, geochronology and palaeoclimatology of the last 15 million years in marine and continental sequences. *Palaeogeography, Palaeoclimatology, Palaeoecology*, **16**, 1–215.

BERGGREN, W. A., KENT, D. V., SWISHER, C. C. III & AUBRY, M.-P. 1995. A revised Cenozoic geochronology and chronostratigraphy. *In*: BERGGREN, W. A., KENT, D. V. AUBRY, M.-P. & HARDENBOL, J. (eds) *Geochronology, time scales and global stratigraphic correlation*. SEPM Special Publications, **54**, 129–212.

BOWEN, D. Q. 1978. *Quaternary Geology*. Pergamon Press, Oxford.

BOWEN, D. Q. (ed.) 1999. *A revised correlation of Quaternary deposits in the British Isles*. Geological Society, London, Special Report **23**.

BOWRING, S. A., ERWIN, D. H., JIN, Y. G., MARTIN, M. W., DAVIDEK, K. & WANG, W. 1998. U–Pb zircon geochronology and tempo of the End-Permian mass extinction. *Science*, **280**, 1039–1045.

CALLOMON, J. H. & CHANDLER, R. B. 1990. A review of the ammonite horizons of the Aalenian–Lower Bajocian Stages in the Middle Jurassic of southern England. *Memoire descrittive della Carta Geologica d'Italia*, **40**, 85–112.

COOPER, R. A. & LINDHOLM, K. 1990. A precise world-wide correlation of early Ordovician graptolite sequences. *Geological Magazine*, **127**, 497–525.

COPE, J. C W. 1993. High resolution biostratigraphy. In: HAILWOOD, E. A. & KIDD, R. B. (eds) *High Resolution Stratigraphy*. Geological Society, London, Special Publications, **70**, 257–265.

COPE, J. C W., GETTY, T. A., HOWARTH, M. K., MORTON, N. & TORRENS, H. S. 1980. *A correlation of Jurassic rocks in the British Isles. Part One: Introduction and Lower Jurassic*. Geological Society, London, Special Report **14**.

COPE, J. C W., INGHAM, J. K. RAWSON, P. F. (eds) 1992. *Atlas of Palaeogeography and Lithofacies*. Geological Society, London, Memoir **13**.

COPESTAKE, P. 1992. Application of micropalaeontology to hydrocarbon exploration in the North Sea Basin. *In*: JENKINS, D. G. (ed.) *Applied Micropalaeontology*. Kluwer Academic, Dordrecht, 93–152.

DICKIN, A. P. 1995. *Radiogenic Isotope Geology*. Cambridge University Press, Cambridge.

EDWARDS, R. A. & SCRIVENER, R. C. 1999. *Geology of the country around Exeter*. Memoir of the British Geological Survey, sheet 325 (England and Wales).

EMERY, D. & MYERS, K. J. 1996. *Sequence Stratigraphy*. Blackwell Scientific, Oxford.

FAURE, G. 1986. *Principles of Isotope Geology*. 2nd edition. John Wiley & Sons Inc., New York.

FRYE, J. C. & WILLMAN, H. B. 1962. Morphostratigraphic units in Pleistocene stratigraphy. *American Association of Petroleum Geologists Bulletin*, **46**, 112–113.

GALLOWAY, W. E. 1989. Genetic Stratigraphic Sequences in Basin Analysis I: Architecture and Genesis of Flooding-Surface Bounded Depositional Units. *American Association of Petroleum Geologists Bulletin*, **734**, 125–142.

GEORGE, T. N., MILLER, T. G., AGER, D. V., BLOW, W. H., CASEY, R., HARLAND, B.,

HOLLAND, C. H., HUGHES, N. F., KELLAWAY, G. A., KENT, P. E., RAMSBOTTOM, W. H. C. &
RHODES, F. H. T. 1967. Report of the stratigraphical code sub-committee. *Proceedings of the
Geological Society, London*, **1638**, 75–87.

GEORGE, T. N., HARLAND, W. B., AGER, D. V., BALL, H. W., BLOW, W. H., CASEY, R.,
HOLLAND, C. H., HUGHES, N. F., KELLAWAY, G. A., KENT, P. E., RAMSBOTTOM, W. H. C.,
STUBBLEFIELD, C. J. & WOODLAND, A. W. 1969. Recommendations on stratigraphical usage.
Proceedings of the Geological Society, London, **1656**, 139–166.

GIBBARD, P. L. 1985. *Pleistocene History of the middle Thames Valley*. Cambridge University Press,
Cambridge.

GIBBARD, P. L. & WEST, R.G. 2000. Quaternary chronostratigraphy: the nomenclature of
terrestrial sequences. *Boreas*, **29**, 329–336.

GREGORY, F. J. 1995. *Middle and Upper Jurassic Foraminifera and Radiolaria of Scotland: An
integerated biostratigraphical and palaeoenvironmental approach*. PhD Thesis, University of Hull.

GREUTER, W., BIRDET, H. M., CHALONER, W. G. ET AL. 1988. *International Code of Botanical
Nomenclature*. Koeltz Scientific Books, Koenigstein, Germany.

HAILWOOD, E. A. 1989. *Magnetostratigraphy*. Geological Society, London, Special Report **19**.

HAQ, B. U., HARDENBOL, J. & VAIL, P. 1988. Mesozoic and Cenozoic chronostratigraphy and
cycles of sea-level change. *In*: WILGUS, C. K., HASTINGS, B. S., KENDALL, G. ST. C.,
POSAMENTIER, H. W., ROSS, C. A. & VAN WAGONER, J. C. (eds) *Sea-Level Changes: an
Integrated Approach*. Society of Economic Paleontologists and Mineralogists, Special Publications, **42**, 71–108.

HARLAND, W. B., AGER, D. V., BALL, H. W., BISHOP, W. W., BLOW, W. H., CURRY, D., DEER,
W. A., GEORGE, T. N., HOLLAND, C. H., HOLMES, S. C. A., HUGHES, N. F., KENT, P. E.,
PITCHER, W. S., RAMSBOTTOM, W. H. C., STUBBLEFIELD, C. J., WALLACE, P. & WOODLAND,
A. W. 1972. A concise guide to stratigraphical procedure. *Journal of the Geological Society,
London*, **128**, 295–305.

HARLAND, W. B., ARMSTRONG, R. L., COX, A. V., CRAIG, L. E., SMITH, A. G. & SMITH, D. G.
1990. *A geologic time scale 1989*. Cambridge University Press, Cambridge.

HAYS, J. D., IMBRIE, J. & SHACKLETON, N. J. 1976. Variations in the Earth's orbit: pacemaker of
the ice ages. *Science*, **194**, 1121–1132.

HEDBERG, H. D. 1976. *International stratigraphic guides: a guide to stratigraphic classification,
terminology and procedure. International Subcommission on stratigraphic classification of IUGS
Commission on Stratigraphy*. Wiley, Chichester.

HELLAND-HANSEN, W. & GJELBERG, J. G. 1994. Conceptual basis and variability in sequence
stratigraphy: a different perspective. *Sedimentary Geology*, **92**, 31–52.

HELLAND-HANSEN, W. & MARTINSON, O. J. 1996. Shoreline trajectories and sequences:
description of variable depositional-dip scenarios. *Journal of Sedimentary Research*, **66**, 670–688.

HESSELBO, S. P. & PARKINSON, D. N. (Eds) 1996. *Sequence Stratigraphy in British Geology*.
Geological Society, London, Special Publication **103**.

HOLLAND, C. H. & BASSETT, M. G. 1988. Discussion on biostratigraphical correlation and the
stages of the Llandovery. *Journal of the Geological Society, London*, **145**, 881–882.

HOLLAND, C. H., AUDLEY-CHARLES, M. G., BASSETT, M. G., COWIE, J. W., CURRY, D., FITCH,
F. J., HANCOCK, J. M., HOUSE, M. R., INGHAM, J. K., KENT, P. E., MORTON, N.,
RAMSBOTTOM, W. H. C., RAWSON, P. F., SMITH, D. B., STUBBLEFIELD, C. J., TORRENS, H.
S., WALLACE, P. & WOODLAND, A. W. 1978. *A guide to stratigraphical procedure*. Geological
Society, London, Special Report **10**.

HOUSE, M. R. & GALE, A. S. (eds) 1995. *Orbital Forcing Timescales and Cyclostratigraphy*.
Geological Society, London, Special Publication **85**.

HUFF, W. D., DAVIS, D., BERGSTROM, S. M., KREKELER, M. P. S., KOLATA, D. R. &
CINGOLANI, C. 1997. A biostratigraphically well-constrained K-bentonite U-Pb age for the
lowermost Darriwilian stage (middle Ordovician) from the Argentine Precordillera. *Episodes*, **20**,
29–33.

HUNT, D. & TUCKER, M. E. 1992. Stranded parasequences and the forced regressive wedge systems
tract: deposition during base-level fall. *Sedimentary Geology*, **81**, 1–9.

HUNT, D. & TUCKER, M. E. 1995. Stranded parasequences and the forced regressive wedge systems tract: deposition during base-level fall – reply. *Sedimentary Geology*, **95**, 147–160.

JEANS, C. V. 1995. Clay mineral stratigraphy in Palaeozoic and Mesozoic red bed facies onshore and offshore UK. *In*: DUNAY, R. E. & HAILWOOD, E.A. (eds) *Non-Biostratigraphical Methods of Dating and Correlation*. Geological Society, London, Special Publications, **89**, 31–55.

KIDWELL, S. M. 1991. Condensed deposits in siliciclastic sequences: expected and observed features. *In*: EINSELE, G., RICKEN, W. & SEILACHER, A. (eds) *Cycles and Events in Stratigraphy*. Berlin, Springer-Verlag, pp. 682–695.

KOWALLIS, B. J. CHRISTIANSEN, E. H., DEINO, A. L., KUNK, M. J. & HEAMAN, L. M. 1995. Age of the Cenomanian–Turonian Boundary in the western Interior of the United States. *Cretaceous Research*, **16**, 109–129.

LANDING, E., BOWRING, S. A., DAVIDEK, K. L., WESTROP, S. R., GEYER, G. & HELDMAIER, W. 1998. Duration of the Early Cambrian: U–Pb ages of volcanic ashes from Avalon and Gondwana. *Canadian Journal of Earth Sciences*, **35**, 329–338.

LEES, A. & MILLER, J. 1985. Facies variation in Waulsortian buildups; part 2, Mid-Dinantian buildups from Europe and North America. *Geological Journal*, **20**, 159–180.

MANGE-RAJETSKY, M. A. 1995. Subdivision and correlation of monotonous sandstone sequences using high-resolution heavy mineral analysis: the Triassic of the Central Graben. *In*: DUNAY, R. E. & HAILWOOD, E. A. (eds) *Non-Biostratigraphical Methods of Dating and Correlation*. Geological Society, London, Special Publication, **89**, 23–30.

MANN, K. O. & LANE, M. R. 1995. *Graphic Correlation*. SEPM Society for Sedimentary Geology, Special Publications **53**.

MARSHALL, J. D. 1992. Climatic and oceanographic isotopic signals from the carbonate rock record and their preservation. *Geological Magazine* **129**, 143–160.

McARTHUR, J. M. 1988. Strontium isotope stratigraphy. *In*: DOYLE, P. & BENNET, M. R. (eds) *Unlocking the Stratigraphical Record*. Wiley, Chichester.

MIALL, A. D. 1997. *The Geology of Stratigraphic Sequences*. Springer-Verlag, Berlin.

MILANKOVITCH, M. 1941. *Kanton der Erdbestrahlung und seine Anwendung auf das Eiszeitenproblem*. Serbian Academy of Sciences, Belgrade, Editions Special, **133**.

MORRISON, R. B. 1985. Pliocene/Quaternary geology, gemomorphology, and tectonics of Arizona. *In*: WEDIE, D. L. (ed.) *Soils and Quaternary Geology of the Southwestern United States*. Geological Society of America, Special Papers, **203**, 123–146.

NEAL, J. E. 1996. A summary of Paleogene sequence stratigraphy in northwest Europe and the North Sea. *In*: KNOX, R. W. O'B., COREFIELD, R. M. & DUNAY, R. E. (eds) *Correlation of the Early Paleogene in Northwest Europe*. Geological Society, London, Special Publications, **101**, 15–42.

NEWELL, N. D. 1967. 'Paraconformities'. *In: Essays in Paleontology and Stratigraphy. R.C. Moore Commemoration Volume*. University of Kansas Department of Geology, Special Publications, **2**, 349–367.

NORTH AMERICAN COMMISSION ON STRATIGRAPHIC NOMENCLATURE 1983. North American Stratigraphic Code. *American Association of Petroleum Geologists Bulletin*, **67**, 841-875.

O'NEILL, B. J., DUVERNAY, A. E. & GEORGE, R. A. 1999. Applied palaeontology: a critical stratigraphical tool in Gulf of Mexico exploration. *In*: JONES, R. W. & SIMMONS, M. D. (eds) *Biostratigraphy in Production and Development Geology*. Geological Society, London, Special Publications, **152**, 243–257.

OLAUSSON, E. (Ed) 1982. *The Pleistocene/Holocene boundary in south-western Sweden*. Sveriges Geologiska Undersökning, C **394**.

OLSEN, P. E. & KENT, D. V. 1996. Milankovitch climate forcing in the tropics of Pangaea during the Late Triassic. *Palaeogeography, Palaeoclimatology, Palaeoecology*, **122**, 1–26.

OPDYKE, N. D. & CHANNELL, J. E. T. 1996. *Magnetic stratigraphy*. Academic Press Inc., San Diego.

PARTINGTON, M. A., COPESTAKE, P., MITCHNER, B. C. & UNDERHILL, J. R. 1993a. Biostratigraphic calibration of genetic stratigraphic sequences in the Jurassic–lowermost Cretaceous (Hettangian–Ryazanian) of the North Sea and adjacent areas. *In*: PARKER, J. R.

(ed.) *Petroleum Geology of North West Europe: Proceedings of the 4th Conference*. Geological Society, London, 371–386.

PARTINGTON, M. A., MITCHNER, B. C., MILTON, N. J. & FRASER, A. J. 1993b. Genetic sequence stratigraphy for the North Sea Late Jurassic and Early Cretaceous: distribution and prediction of Kimmeridgian–Late Ryazanian reservoirs in the North Sea and adjacent areas. *In*: PARKER, J. R. (ed.) *Petroleum Geology of North West Europe: Proceedings of the 4th Conference*. Geological Society, London, 347–370.

PEARCE, T. J. & JARVIS, I. 1995. High-resolution chemostratigraphy of Quaternary distal turbidites: a case study of new methods for the analysis and correlation of barren sequences. *In*: DUNAY, R. E. & HAILWOOD, E. A. (eds) *Non-biostratigraphical methods of Dating and Correlation*. Geological Society, London, Special Publication, **89**, 23–30.

PENCK, A. & BRÜCKNER, E. 1909–11. *Die Alpen im Eiszeitalter*. Tauchnitz, Leipzig.

POSAMENTIER, H. W. & VAIL, P. R. 1988. Eustatic controls on clastic deposition II - sequence and systems tract models. *In*: WILGUS, C. K., HASTINGS, B. S., KENDALL, G. St. C., POSAMENTIER, H. W., ROSS, C. A. & VAN WAGONER, J. C. (eds) *Sea-Level Changes: an Integrated Approach*. Society of Economic Paleontologists and Mineralogists, Special Publications **42**, 125–154.

PRESTON, J., HARTLEY, A., HOLE, M., BUCK, S., BOND, J., MANGE, M. & STILL, J. 1998. Integrated whole-rock trace element geochemistry and heavy mineral chemistry studies: aids to the correlation of continental red-bed reservoirs in the Beryl Field, UK North Sea. *Petroleum Geoscience*, **4**, 7–16.

RAMSBOTTOM, W. H. C. 1977. Major cycles of transgression and regression (mesothems) in the Namurian. *Proceedings of the Yorkshire Geological Society*, **41**, 261–291.

RAWSON, P. F., ALLEN, P. & GALE, A. S. 2001. The Chalk Group – a revised lithostratigraphy. *Geoscientist*, **11**(1), 21.

RAWSON, P. F & RILEY, L. A. 1982. Latest Jurassic - Early Cretaceous Events and the "Late Cimmerian Unconformity" in North Sea Area. *American Association of Petroleum Geologists Bulletin*, **66**, 2628–2648.

RAWSON, P. F. & WRIGHT, J. K. 2000. *The Yorkshire Coast*. Geologists' Association Guide **34**.

REMANE, J., BASSETT, M. G., COWIE, J. W., GOHRBANDT, K. H., LANE, H. R., MICHELSEN, O. & NAIWEN, W. 1996. Revised guidelines for the establishment of global chronostratigraphic standards by the International Commission on Stratigraphy. *Episodes*, **19**, 77–81.

RICHARDS, J. P. & NOBLE, S. R. 1998. Application of radiogenic isotope systematics to the timing and origin of hydrothermal systems. *In*: RICHARDS, J. P. & LARSEN, P. B. (eds) *Techniques in hydrothermal ore deposits geology*. Reviews in Economic Geology, **10**, 195–233.

RICHMOND, G. M. 1959. Report of the Pleistocene committee, American Commission on stratigraphic nomenclature. *American Association of Petroleum Geologists, Bulletin*, **43**, 633–675.

RIDE, W. D. L., SABROSKY, C. W., BERNARDI, G. & MELVILLE, R. V. 1985. *International Code of Zoological Nomenclature*. International Trust for Zoological Nomenclature, British Museum (Natural History) London. University of California Press.

RIDER, M. 1998. *Geological interpretation of well logs*. 2nd edition. Whittles Publishers.

ROTH, E. & POTY, B. (eds) 1989. *Nuclear Methods of Dating*. Kluwer Academic Publishers.

RUSHTON, A. W. A. & HOWELLS, M. F. 1998. *Stratigraphical Framework for the Ordovician of Snowdonia and the Lleyn Peninsula*. British Geological Survey Research Report, **RR/98/1** (second edition **RR/99/08**).

SALVADOR, A. 1994. *International Stratigraphic Guide. A Guide to Stratigraphic classification, terminology, and Procedure*. 2nd edition. The International Union of Geological Sciences and the Geological Society of America.

SCHLAGER, W. 1989. Drowning unconformities on carbonate platforms. *In*: CREVELLO, P. D., WILSON, J. L., SARG, J. F. & READ, J. F. (eds) *Controls on Carbonate Platform and Basin Development*. Society of Economic Paleontologists and Mineralogists, Special Publications, **44**, 15–25.

SCHLAGER, W. 1992. *Sedimentology and Sequence Stratigraphy of Reefs and Carbonate Platforms*. American Association of Petroleum Geologists, Continuing Education Course Note Series 34.

SHACKLETON, N. J., McCAVE, I. N. & WEEDON, G. P. (eds) 1999. Astronomical (Milankovitch) calibration of the geological time-scale. *Philosophical Transactions of the Royal Society*, **A357**, 1731–2007.

SHAW, A. B. 1964. *Time in Stratigraphy*. McGraw-Hill, New York.

SHIPP, D. 1999. Wellsite biostratigraphy of Danish horizontal wells. *In*: JONES, R. W. & SIMMONS, M. D. (eds) *Biostratigraphy in Production and Development Geology*. Geological Society, London, Special Publications, **152**, 75-84.

SHUEY, R. T., BROWN, F. H., ECK, G. G. & HOWELL, F.C. 1978. A statistical approach to temporal biostratigraphy. *In*: BISHOP, W. W. (ed.) *Geological background to Fossil Man. Recent Research in the Gregory Rift Valley, East Africa*. Geological Society, London, [Special Publications, **6**], 103–124.

SPEARS, D.A., KANARIS-SOTIRIOU, R., RILEY, N. & KRAUSE, P. 1999. Namurian bentonites in the Pennine Basin, UK – origin and magmatic affinities. *Sedimentology*, **46**, 385–401.

STONE, P. 1995. *Geology of the Rhins of Galloway district*. Memoir of the British Geological Survey, sheets 1 and 3 (Scotland).

SUGGATE, R. P. & WEST, R. G. 1969. *Stratigraphic nomenclature and subdivision in the Quaternary*. Working Group for Stratigraphic Nomenclature, INQUA Commission for Stratigraphy (unpublished discussion document).

TUCKER, R. D., BRADLEY, D. C., STRAETEN, C. A. V., HARRIS, A. G., EBERT, J. R. & MCCUTCHEON, S.R. 1998. New U–Pb zircon ages and the duration and division of Devonian time. *Earth and Planetary Science Letters*, **158**, 175–186.

TURNER, C. 1998. Volcanic maars, long Quaternary sequences and the work of the INQUA Subcommission on European Quaternary Stratigraphy. *Quaternary International*, **47/48**, 41–49.

VAIL, P. R., AUDEMARD, F., BOWMAN, S. A., EISNER, P. N. & PEREZ-CRUZ, C. 1991. The stratigraphic signatures of tectonics, eustacy and sedimentology - an overview. *In*: EINSELE, G., RICKEN, W. & SEILACHER, A. (eds) *Cycles and Events in Stratigraphy*. Springer-Verlag, Berlin, 617–659.

VAN WAGONER, J. C., MITCHUM, R. M., CAMPION, K. M. & RAHMANIAN, V. D. 1990. *Siliciclastic sequence stratigraphy in well logs, cores, and outcrops*. American Association of Petroleum Geologists, Methods in Exploration Series, **7**.

VAN WAGONER, J. C., POSAMENTIER, H.W., MITCHUM, R. M., VAIL, P. R., SARG, J. F., LOUTIT, T. S. & HARDENBOL, R. J. 1988. An overview of the fundamentals of sequence stratigraphy and key definitions. *In*: WILGUS, C. K., HASTINGS, B. S., KENDALL, G. St. C., POSAMENTIER, H. W., ROSS, C. A. & VAN WAGONER, J. C. (eds) *Sea-Level Changes: an Integrated Approach*. Society of Economic Paleontologists and Mineralogists, Special Publications, **42**, 39–45.

VOLLSET, J. & DORÉ, A. G. (eds) 1984. *A revised Triassic and Jurassic lithostratigraphic nomenclature for the Norwegian North Sea*. Bulletin of the Norwegian Petroleum Directorate **3**.

WHITTAKER, A., COPE, J. C. W., COWIE, J. W., GIBBONS, W., HAILWOOD, E. A., HOUSE, M. R., JENKINS, D. G., RAWSON, P. F., RUSHTON, A. W. A., SMITH, D. G., THOMAS, A. T. & WIMBLEDON, W. A. 1991. A guide to stratigraphical procedure. *Journal of the Geological Society, London*, **148**, 813-824. [Reprinted 1992 as Geological Society, Special Report, **20**].

WHITTAKER, A., HOLLIDAY, D. W. & PENN, I. E. 1985. *Geophysical logs in British stratigraphy*. Geological Society, London, Special Report **18**.

WHITTEN, E. H. T. 1991. Granitoid suites. *Geological Journal*, **26**, 117–122.

WOOD, C. J. & SMITH, E. G. 1978. Lithostratigraphical nomenclature of the Chalk in North Yorkshire, Humberside and Lincolnshire. *Proceedings of the Yorkshire Geological Society*, **42**, 263–287.

WOODCOCK, N. H. 1990. Sequence stratigraphy of the Palaeozoic Welsh Basin. *Journal of the Geological Society, London*, **147**, 537–547.

WRAY, D. S. & GALE, A. S. 1993. Geochemical correlation of marl bands in Turonian chalks of the Anglo-Paris Basin. *In*: HAILWOOD, E. A. & KIDD, R. B. (eds) *High Resolution Stratigraphy*. Geological Society, London, Special Publications, **70**, 211–226.

Index

Page numbers in *italics* refer to Figures.

Special Reports of the Geological Society

No. 21: *A revised correlation of the Silurian Rocks in the British Isles*
By L. R. M. Cocks, C. H. Holland & R. B. Rickards
ISBN: 0-903317-77-3, 32 pages, £14/$24

No. 22: *A revised correlation of the Precambrian Rocks of the British Isles*
Edited by W. Gibbons & A. L. Harris
ISBN: 1-897799-11-X, 112 pages, £14/$22

No. 23: *A revised correlation of Quaternary Deposits in the British Isles*
Edited by D. Q. Bowen
ISBN: 1-86239-042-8, 176 pages, £39/$65

No. 24: *A revised correlation of the Ordovician Rocks in the British Isles*
By R. A. Fortey *et al.*
ISBN: 1-86239-069-X, 88 pages, £18/$30

Revised correlations of other periods are in preparation.

Published by The Geological Society of London from:
The Geological Society Publishing House
Unit 7, Brassmill Enterprise Centre
Brassmill Lane
Bath BA1 3JN, UK

Orders: Tel. +44 (0)1225 445046
           Fax +44 (0)1225 442836
Online bookshop: http://bookshop.geolsoc.org.uk